HIGH SCHOOL ALGEBRA I

A Comprehensive Review and
Step-by-Step Guide to Mastering
Algebra 1

By

Reza Nazari

Copyright © 2024
Effortless Math Education

All rights reserved. No part of this publication may be reproduced, stored in a retrieval system, or transmitted in any form or by any means, electronic, mechanical, photocopying, recording, scanning, or otherwise, except as permitted under Section 107 or 108 of the 1976 United States Copyright Act, without permission of the author.

All inquiries should be addressed to:
info@effortlessMath.com
www.EffortlessMath.com

Published by: **Effortless Math Education**

for Online Math Practice Visit www.EffortlessMath.com

Welcome to
High School Algebra I Prep 2024

Thank you for choosing Effortless Math for your High School Algebra I preparation and congratulations on making the decision to take the High School Algebra I course! It's a remarkable move you are taking, one that shouldn't be diminished in any capacity.

That's why you need to use every tool possible to ensure you succeed on the test with the highest possible score, and this extensive study guide is one such tool.

High School Algebra I is designed to be comprehensive and cover all the topics that are typically covered in a High School Algebra I course. It provides clear explanations and examples of the concepts and includes practice problems and quizzes to test your understanding of the material. The textbook also provides step-by-step solutions to the problems, so you can check your work and understand how to solve similar problems on your own.

Additionally, this textbook is written in a user-friendly way, making it easy to follow and understand even if you have struggled with math in the past. It also includes a variety of visual aids such as diagrams, graphs, and charts to help you better understand the concepts.

High School Algebra I is flexible and can be used to supplement a traditional classroom setting, or as a standalone resource for self-study. With the help of this comprehensive textbook, you will have the necessary foundation to master the material and succeed in the High School Algebra I course.

Effortless Math's High School Algebra I Online Center

Effortless Math Online High School Algebra I Center offers a complete study program, including the following:

- ✓ Step-by-step instructions on how to prepare for the Algebra I test

- ✓ Numerous Algebra I worksheets to help you measure your math skills

- ✓ Video lessons for all High School Algebra I topics

- ✓ Full-length Algebra I practice tests

- ✓ And much more…

No Registration Required.

Visit **EffortlessMath.com/Algebra1** to find your online Algebra I resources.

How to Use This Book Effectively?

Look no further when you need a study guide to improve your math skills to succeed on the High School Algebra I course. Each chapter of this comprehensive guide to the Algebra I will provide you with the knowledge, tools, and understanding needed for every topic covered on the course.

It's imperative that you understand each topic before moving onto another one, as that's the way to guarantee your success. Each chapter provides you with examples and a step-by-step guide of every concept to better understand the content that will be on the course. To get the best possible results from this book:

- **Begin studying long before your test date.** This provides you ample time to learn the different math concepts. The earlier you begin studying for the test, the sharper your skills will be. Do not procrastinate! Provide yourself with plenty of time to learn the concepts and feel comfortable that you understand them when your test date arrives.
- **Practice consistently.** Study Algebra I concepts at least 20 to 30 minutes a day. Remember, slow and steady wins the race, which can be applied to preparing for the Algebra I test. Instead of cramming to tackle everything at once, be patient and learn the math topics in short bursts.
- Whenever you get a math problem wrong, **mark it off, and review it later** to make sure you understand the concept.
- Start each session by **looking over the previous material.**
- Once you've reviewed the book's lessons, **take a practice test at the back of the book** to gauge your level of readiness. Then, review your results. Read detailed answers and solutions for each question you missed.
- **Take another practice test** to get an idea of how ready you are to take the actual exam. Taking the practice tests will give you the confidence you need on test day. Simulate the Algebra I testing environment by sitting in a quiet room free from distraction. Make sure to clock yourself with a timer.

Looking for more?

Visit EffortlessMath.com/Algebra1 to find hundreds of Algebra I worksheets, video tutorials, practice tests, High School Algebra I formulas, and much more.

Or scan this QR code.

No Registration Required.

Contents

Chapter: Integers and Order of Operations — 1

1
- Adding and Subtracting Integers 2
- Multiplying and Dividing Integers 4
- Translate a Phrase into an Algebraic Statement 6
- Order of Operations 8
- Integers and Absolute Value 10
- Proportional Ratios 12
- Similarity and Ratios 14
- Percent Problems 16
- Percent of Increase and Decrease 18
- Discount, Tax, and Tip 20
- Simple Interest 22
- The Distributive Property 24
- Approximating Irrational Numbers 26
- Chapter 1: Answers 28

Chapter: Exponents and Variables — 31

2
- Multiplication Property of Exponents 32
- Division Property of Exponents 34
- Powers of Products and Quotients 36
- Zero and Negative Exponents 38
- Negative Exponents and Negative Bases 40
- Match Exponential Functions and Graphs 42
- Write Exponential Functions: Word Problems 44
- Scientific Notation 46
- Addition and Subtraction in Scientific Notation 48
- Multiplication and Division in Scientific Notation 50
- Chapter 2: Answers 52

Chapter: Expressions and Equations — 53

3
- Simplifying Variable Expressions 54
- Evaluating One Variable 56
- Evaluating Two Variables 58
- One–Step Equations 60
- Multi–Step Equations 62
- Rearrange Multi-Variable Equations 64
- Finding Midpoint 66
- Finding the Distance of Two Points 68
- Graphing Absolute Value Equations 70
- Chapter 3: Answers 72

EffortlessMath.com

Chapter 4: Linear Functions — 75

- Finding Slope 76
- Writing Linear Equations 78
- Graphing Linear Inequalities 80
- Finding the Slop, x –intercept and y –intercept 82
- Write an Equation from a Graph 84
- Slope-intercept Form and Point-slope Form 86
- Write a Point-slope Form Equation from a Graph 88
- Find x – and y –intercepts in the Standard Form of Equation 90
- Graph an Equation in the Standard Form 92
- Equations of Horizontal and Vertical Lines 94
- Graph a Horizontal or Vertical Line 96
- Graph an Equation in Point-Slope Form 98
- Equation of Parallel and Perpendicular Lines 100
- Compare Linear Function's Graph and Equations 102
- Two-variable Linear Equations Word Problems 104
- Chapter 4: Answers 106

Chapter 5: Inequalities and System of Equations — 109

- One–Step Inequalities 110
- Multi–Step Inequalities 112
- Compound Inequalities 114
- Write a Linear Inequality from a Graph 116
- Graph Solutions to One-step and Two-step Linear Inequalities 118
- Solve Advanced Linear Inequalities in Two-Variables 120
- Graph Solutions to Advance Linear Inequalities 122
- Absolute Value Inequalities 124
- System of Equations 126
- Find the Number of Solutions to a Linear Equation 128
- Write a System of Equations Given a Graph 130
- Systems of Equations Word Problems Top of Form 132
- Solve One-step and Two-step Linear Equations Word Problems 134
- Systems of Linear Inequalities 136
- Write Two-variable Inequalities in Form of Word Problems 138
- Chapter 5: Answers 140

Chapter 6: Quadratic — 143

- Solving Quadratic Equations 144
- Graphing Quadratic Functions 146
- Solve a Quadratic Equation by Factoring 148
- Transformations of Quadratic Functions 150
- Quadratic Formula and the Discriminant 152
- Characteristics of Quadratic Functions: Equations 154
- Characteristics of Quadratic Functions: Graphs 156
- Complete a Function Table: Quadratic Functions 158
- Domain and Range of Quadratic Functions: Equations 160

Contents

Chapter 6

- Factor Quadratics: Special Cases 162
- Factor Quadratics Using Algebra Tiles 164
- Write a Quadratic Function from Its Vertex and Another Point 166
- Chapter 6: Answers 168

Chapter 7: Polynomials — 171

- Simplifying Polynomials 172
- Adding and Subtracting Polynomials 174
- Add and Subtract Polynomials Using Algebra Tiles 176
- Multiplying Monomials 178
- Dividing Monomials 180
- Multiplying a Polynomial and a Monomial 182
- Multiply Polynomials Using Area Models 184
- Multiplying Binomials 186
- Multiply two Binomials Using Algebra Tiles 188
- Factoring Trinomials 190
- Factoring Polynomials 192
- Use a Graph to Factor Polynomials 194
- Factoring Special Case Polynomials 196
- Add Polynomials to Find Perimeter 198
- Chapter 7: Answers 200

Chapter 8: Relations and Functions — 203

- Function Notation and Evaluation 204
- Adding and Subtracting Functions 206
- Multiplying and Dividing Functions 208
- Composition of Functions 210
- Evaluate an Exponential Function 212
- Function Inverses 214
- Domain and Range of Relations 216
- Interval Notation 218
- Rate of Change and Slope 220
- Complete a Function Table from an Equation 222
- Chapter 8: Answers 224

Chapter 9: Radical Expressions — 227

- Simplifying Radical Expressions 228
- Adding and Subtracting Radical Expressions 230
- Multiplying Radical Expressions 232
- Rationalizing Radical Expressions 234
- Radical Equations 236
- Domain and Range of Radical Functions 238
- Simplify Radicals with Fractions 240
- Chapter 9: Answers 242

Contents

Chapter 10: Rational Expressions — 243

- Simplifying Complex Fractions 244
- Graphing Rational Functions 246
- Adding and Subtracting Rational Expressions...248
- Multiplying Rational Expressions...250
- Dividing Rational Expressions 252
- Evaluate Integers Raised to Rational Exponents.....................................254
- Chapter 10: Answers 256

Chapter 11: Statistics and Probabilities — 259

- Mean, Median, Mode, and Range of the Given Data...............................260
- Pie Graph..262
- Scatter Plots 264
- Probability Problems..266
- Permutations and Combinations 268
- Calculate and Interpret Correlation Coefficients.....................................270
- Find the Equation of a Regression Line and Interpret Regression Lines.272
- Correlation and Causation..274
- Chapter 11: Answers 276

Chapter 12: Direct and Inverse Variation — 277

- Find the Constant of Variation...278
- Model Inverse Variation 280
- Write and Solve Direct Variation Equations 282
- Chapter 12: Answers 284

Chapter 13: Number Sequences — 285

- Evaluate Recursive Formulas for Sequences...286
- Evaluate Variable Expressions for Number Sequences...........................288
- Write Variable Expressions for Arithmetic Sequences 290
- Write Variable Expressions for Geometric Sequences 292
- Write a Formula for a Recursive Sequence 294
- Chapter 13: Answers 296

High School Algebra I Practice Test 1 — 299

High School Algebra I Practice Test 2 — 323

High School Algebra I Practice Tests Answer Keys — 344

High School Algebra I Practice Tests 1 Explanations — 345

High School Algebra I Practice Tests 2 Explanations — 363

CHAPTER 1
Integers and Order of Operations

Math topics that you'll learn in this chapter:

- ☑ Adding and Subtracting Integers
- ☑ Multiplying and Dividing Integers
- ☑ Translate a Phrase into an Algebraic Statement
- ☑ Order of Operations
- ☑ Integers and Absolute Value
- ☑ Proportional Ratios
- ☑ Similarity and Ratios
- ☑ Percent Problems
- ☑ Percent of Increase and Decrease
- ☑ Discount, Tax, and Tip
- ☑ Simple Interest
- ☑ The Distributive Property
- ☑ Approximating Irrational Numbers

Adding and Subtracting Integers

Integers are a set of numbers that include zero, counting numbers, and the negative versions of counting numbers. They can be represented on a number line as $\{\ldots,-3,-2,-1,0,1,2,3,\ldots\}$.

To add a positive integer, move to the right on the number line, which will result in a larger number.

To add a negative integer, move to the left on the number line, which will result in a smaller number. To subtract an integer, add its opposite.

Number line

Adding integers: When adding integers, the signs of the numbers being added determine the overall sign of the result.

- If both numbers have the same sign (both positive or both negative), add their absolute values and give the result the same sign as the numbers being added.

- If the numbers have different signs (one positive and one negative), subtract the absolute value of the smaller number from the absolute value of the larger number and give the result the same sign as the larger number.

Let's review some examples of adding integers:

- $2 + 3 = 5$ (Both numbers are positive)

- $-4 + (-2) = -6$ (Both numbers are negative)

- $5 + (-3) = 2$ (5 is positive and -3 is negative, so we subtract the absolute value of -3 from 5 and the result is positive)

Integers and Order of Operations

Subtracting integers: When subtracting integers, it is the same as adding the opposite of the number being subtracted. To subtract a positive number, add its opposite (negative) To subtract a negative number, add its opposite (positive)

Let's review some examples of subtracting integers:

- $5 - 2 = 3$ (Same as $5 + (-2)$)
- $-2 - (-5) = 3$ (Same as $-2 + 5$)
- $7 - (-3) = 10$ (Same as $7 + 3$)

Consider another example: Solve. $6 + (3 - 12) =$

First, subtract the numbers in brackets, $3 - 12 = -9$.

Then: $6 + (-9) = \rightarrow$ change addition into subtraction: $6 - 9 = -3$.

Let's do another example together: Solve. $(-4) - (-7) =$

Keep the first number and convert the sign of the second number to its opposite. Change subtraction into addition.

Then: $(-4) + 7 = 3$.

✎ Find each sum or difference.

1) $9 + (-4) =$

2) $(-2 + 6) - (1 - 5) =$

3) $-7 + 9 =$

4) $4 + (-4 - 2) =$

5) $(-4 + 2) - 8 =$

6) $(-14) - (-19) =$

7) $2 + (-3 + 2) - 1 =$

8) $3 - (-11 - 3) =$

9) $(-7 + 9) - 39 =$

10) $(8 - 19) - (-4 + 12) =$

Multiplying and Dividing Integers

When multiplying and dividing integers, the signs of the numbers being multiplied determine the overall sign of the result.

- If both numbers are positive, the result will be positive.
- If both numbers are negative, the result will be positive.
- If one number is positive and the other is negative, the result will be negative.

Use the following rules for multiplying and dividing integers:

- $(negative) \times (negative) = positive$
- $(negative) \div (negative) = positive$
- $(negative) \times (positive) = negative$
- $(negative) \div (positive) = negative$
- $(positive) \times (positive) = positive$
- $(positive) \div (negative) = negative$

Multiplying integers: When multiplying integers, the signs of the numbers being multiplied determine the overall sign of the result.

- If both numbers are positive, the result will be positive.
- If both numbers are negative, the result will be positive.
- If one number is positive and the other is negative, the result will be negative.

Let's review some examples of multiplying integers:

- $2 \times 3 = 6$ (Both numbers are positive)
- $-2 \times (-5) = 10$ (Both numbers are negative)
- $2 \times (-5) = -10$ (One number is positive, the other is negative)

Integers and Order of Operations

Dividing integers: When dividing integers, the signs of the numbers being divided determine the overall sign of the result.

- If both numbers are positive, the result will be positive.
- If both numbers are negative, the result will be positive.
- If the dividend (the number being divided) is positive and the divisor (the number doing the dividing) is negative, the result will be negative.
- If the dividend is negative and the divisor is positive, the result will be negative.

Dividing integers:

- $\frac{8}{4} = 2$ (Both numbers are positive)
- $\frac{-8}{-4} = 2$ (Both numbers are negative)
- $\frac{8}{-4} = -2$ (The dividend is positive; the divisor is negative)
- $\frac{-8}{4} = -2$ (The dividend is negative; the divisor is positive)

It's important to note that when dividing integers if the result is not a whole number, it will be expressed as a fraction or decimal. It's also important to pay attention to the rules of the algebra and not mix the signs and operations.

✎ Solve.

11) $(3 - 9) \div (-3) =$

12) $-25 \times (-3 + 1) =$

13) $(-4) \times 3 \times (-2) =$

14) $(-49) \div (-7) =$

15) $2 \times (-5) =$

16) $(-5) \times 6 =$

17) $(-63) \div (15 - 6) =$

18) $(-32) \times (-1) \div (-4) =$

19) $(-15) \div (-17 + 12) =$

20) $(-64) \div (-16 + 8) =$

Translate a Phrase into an Algebraic Statement

An algebraic statement is a mathematical sentence that uses variables and operations (such as $+, -, \times, \div$) to express a relationship or equality. To translate a phrase into an algebraic statement, you can identify the variables and operations in the phrase and write them out using mathematical symbols.

Let's review some examples:

- The phrase "the product of a number and 3 is 12" can be translated into the algebraic statement:

$$a \times 3 = 12$$

Where "a" is the variable representing the number, and "$\times$" represents multiplication, and "$=$" represents equality.

- The phrase "the sum of a number and 4 is equal to 10" can be translated into the algebraic statement:

$$a + 4 = 10$$

- Where "a" is the variable representing the number, and "$+$" represents addition, and "$=$" represents equality.

In general, phrases like "the difference between two numbers" can be translated into an algebraic statement using the subtraction operator "$-$" and "the quotient of two numbers" can be translated into an algebraic statement using the division operator "$\div$".

Translating keywords and phrases into algebraic expressions:

- Addition: plus, more than, the sum of, etc.
- Subtraction: minus, less than, decreased, etc.
- Multiplication: times, product, multiplied, etc.
- Division: quotient, divided, ratio, etc.

EffortlessMath.com

To better understand the issue, let's translate some expressions into algebraic expressions:

8 more than a number is 45.

More than mean plus a number $+x$.

Then: $8 + x = 45$.

17 times the sum of 3 and x.

The sum of 3 and x: $3 + x$. Times means multiplication. Then: $17 \times (3 + x) = 17(3 + x)$.

Three more than a number is 19.

"More than" means plus. "a number" is x.

Then: $3 + x = 19$.

16 times the sum of 5 and x.

The sum of 5 and x: $5 + x$. Times means multiplication.

Then: $16 \times (5 + x) = 16(5 + x)$.

Seven more than a number is 28.

"More than" means plus. "a number" is x.

Then: $7 + x = 28$.

25 times the sum of 1 and x.

The sum of 1 and x: $1 + x$. Times means multiplication.

Then: $25 \times (1 + x) = 25(1 + x)$.

✎ Write an algebraic expression for each phrase.

21) -3 divided by a is 18.

22) The difference between 10 and x is 9.

23) The cube of 4.

24) The square of b is 25.

25) -5 more than a number is 7.

26) The sum of x and 1 times 5 is 25.

Order of Operations

The order of operations is a set of rules that dictate the order in which mathematical operations should be performed. This ensures that all mathematical expressions are evaluated consistently and correctly. The acronym PEMDAS is often used to help remember the order of operations:

P: Parentheses

E: Exponents (i.e. powers and square roots, etc.)

MD: Multiplication and Division (from left to right)

AS: Addition and Subtraction (from left to right)

When working with an algebraic equation, it's important to remember to follow the order of operations to simplify the equation correctly. Parentheses must be simplified first, followed by exponents, then any multiplication and division, and finally any addition and subtraction.

For example: $3 + 4 \times 5 = 3 + 20 = 23$.

If one doesn't follow the order of operations, the result would be incorrect:

$$7 \times 5 = 35$$

It's also important to remember that when there are multiple operations of the same level of precedence, you should work from left to right.

For example: $4 \times 5 \div 2 + 3 = 20 \div 2 + 3 = 10 + 3 = 13$.

It's important to follow the order of operations to ensure that mathematical expressions are evaluated consistently and correctly.

Let's review an example:

Calculate. $(2 + 3) \times (4 - 7) + 5$

1. First, we simplify anything inside of parentheses. So, we have: (2 + 3) = 5 and (4 − 7) = −3.

2. Next, we simplify any exponents, but there are none in this case.

3. Then, we simplify any multiplication and division, working from left to right. We have: 5 × (−3) = −15.

4. Finally, we simplify any addition and subtraction, working from left to right. We have: −15 + 5 = −10.

So, the final simplified expression is −10.

Another example:

Calculate. −5[(5 × 8) ÷ (10 × 2)] =

First, calculate within parentheses:

−5[(40) ÷ (10 × 2)] = −5[(40) ÷ (20)] = −5[2].

Multiply −5 and 2. Then: −5[2] = −10.

Another example for better understanding:

Solve. (63 ÷ 9) + (−15 + 12) =

First, calculate within parentheses:

(63 ÷ 9) + (−15 + 12) = (7) + (−3).

Then: (7) − (3) = 4.

✏ Evaluate each expression.

27) (7 − 3) − (28 ÷ 7) =

28) [(2 × 3) + (1 − 3)] × (−1) =

29) (−3) × 2 + 5 =

30) (−11 + 6) + (2 × 4) =

31) (−9) + (4 × 3) =

32) (−2 + 11) − (15 ÷ 5) =

33) [(63 ÷ 9) + 17] ÷ 3 − 18 =

34) −2(15 − 4) − (−24) =

Integers and Absolute Value

Absolute value is the distance a number is from zero on the number line, regardless of its sign. It is represented by two vertical lines (| |) surrounding the number. For example, the absolute value of −5 is 5 and the absolute value of 5 is also 5. Absolute value can be thought of as the magnitude of a number, regardless of its sign.

The absolute value of any real number is always non-negative, so $|a| \geq 0$.

$|-5| = 5$ (The absolute value of −5 is 5) $|5| = 5$ (the absolute value of 5 is also 5)

$|-8| = 8$ (the absolute value of −8 is 8) $|0| = 0$ (the absolute value of 0 is 0).

You can use absolute value to find the distance between two numbers on a number line, regardless of their order. For example, $|3 - 6| = |-3| = 3$, which is the distance between the numbers 3 and 6 on the number line.

Let's review some examples:

1. Calculate. $|14 - 2| \times 5 =$

First, solve $|14 - 2|$: → $|14 - 2| = |12|$, the absolute value of 12 is 12, $|12| = 12$.

Then: $12 \times 5 = 60$.

2. Solve. $\frac{|-24|}{4} \times |5 - 7| =$

First, find $|-24|$ → the absolute value of −24 is 24. Then: $|-24| = 24$, $\frac{24}{4} \times |5 - 7| =$.

Now, calculate $|5 - 7|$, → $|5 - 7| = |-2|$, the absolute value of −2 is 2: $|-2| = 2$.

Then: $\frac{24}{4} \times 2 = 6 \times 2 = 12$.

Integers and Order of Operations

3. Solve. $|8 - 2| \times \frac{|-4 \times 7|}{2} =$

First, calculate $|8 - 2|$, → $|8 - 2| = |6|$, the absolute value of 6 is 6, $|6| = 6$. Then: $6 \times \frac{|-4 \times 7|}{2}$.

Now calculate $|-4 \times 7|$, → $|-4 \times 7| = |-28|$, the absolute value of -28 is 28, $|-28| = 28$. Then: $6 \times \frac{28}{2} = 6 \times 14 = 84$.

4. Solve. $\frac{|-36|}{4} - |20 - 44| =$

First, find $|-36|$ → the absolute value of -36 is 36. Then: $|-36| = 36$, $\frac{36}{4} - |20 - 44| =$.

Now, calculate $|20 - 44|$, → $|20 - 44| = |-24|$, the absolute value of -24 is 24: $|-24| = 24$. Then: $\frac{36}{4} - 24 = 9 - 24 = -15$.

5. Solve. $|12 - 9| \times \frac{|-32 \div 8|}{4} =$

First, calculate $|12 - 9|$, → $|12 - 9| = |3|$, the absolute value of 3 is 3, $|3| = 3$. Then: $3 \times \frac{|-32 \div 8|}{4}$.

Now calculate $|-32 \div 8|$, → $|-32 \div 8| = |-4|$, the absolute value of -4 is 4, $|-4| = 4$. Then: $3 \times \frac{4}{4} = 3 \times 1 = 3$.

✏ Find the answers.

35) $\frac{|3 \times (-10)|}{5} \div \frac{|-21|}{7} =$

36) $|2 - 8| - |-11| =$

37) $|-4| - |8 + 1| =$

38) $-3 + |8 - 2| - |-3 + 2| =$

39) $\frac{|-21|}{7} + |-3 + 4| =$

40) $|-2 - 3| + |1 + 4| =$

41) $-|11 - 3| + |7| =$

42) $|-5 + 9| \times |-2| =$

Proportional Ratios

Two ratios are proportional if they represent the same relationship between the quantities they compare. For example, if the ratio of length to width in a rectangle is always $2:1$, then the rectangle has a proportional ratio of $2:1$. This means that no matter the specific measurements of the rectangle, the ratio of length to width will always be $2:1$.

A proportion means that two ratios are equal. It can be written in two ways:

Using the symbol "=" to indicate equality: $\frac{a}{b} = \frac{c}{d}$.

Using the word "is" or "to" to indicate equality: a is to b as c is to d.

For example, if the ratio of length to width in a rectangle is $2:1$, this can be written as a proportion: $\frac{2}{1} = \frac{\text{length}}{\text{width}}$, or 2 is to 1 as length is to width.

The proportion $\frac{a}{b} = \frac{c}{d}$ can be written as: $a \times d = c \times b$.

Let's review this concept with an example:

Solve this proportion for x. $\frac{2}{5} = \frac{6}{x}$

Use cross multiplication:

$\frac{2}{5} = \frac{6}{x} \rightarrow 2 \times x = 6 \times 5 \rightarrow 2x = 30$.

Divide both sides by 2 to find x:

$x = \frac{30}{2} \rightarrow x = 15$.

Another example:

If the ratio of apples to oranges in a basket is $3:2$, how many oranges are there if there are 12 apples?

EffortlessMath.com

To find the number of oranges, we can use the proportion $3:2 = 12:x$ (where x represents the number of oranges). To solve for x, we can cross-multiply and divide: $3x = 24, x = 8$. So, there are 8 oranges in the basket.

Another example for Proportional Ratios:

Solve this proportion for x. $\frac{5}{7} = \frac{20}{x}$

Use cross multiplication:

$\frac{5}{7} = \frac{20}{x} \rightarrow 5 \times x = 7 \times 20 \rightarrow 5x = 140$.

Divide to find x:

$x = \frac{140}{5} \rightarrow x = 28$.

✎ Solve each proportion.

43) $\frac{3}{5} = \frac{6}{x}$, $x =$

44) $\frac{x}{9} = \frac{4}{27}$, $x =$

45) $\frac{2}{a} = \frac{18}{3}$, $a =$

46) $\frac{y}{35} = \frac{2}{5}$, $y =$

47) $\frac{27}{6} = \frac{x}{2}$, $x =$

48) $\frac{1}{8} = \frac{7}{w}$, $w =$

49) $\frac{7}{x} = \frac{2}{6}$, $x =$

50) $\frac{2}{x} = \frac{4}{10}$, $x =$

51) $\frac{3}{2} = \frac{x}{8}$, $x =$

52) $\frac{x}{6} = \frac{5}{3}$, $x =$

53) $\frac{3}{9} = \frac{5}{x}$, $x =$

54) $\frac{4}{18} = \frac{2}{x}$, $x =$

55) $\frac{6}{16} = \frac{3}{x}$, $x =$

56) $\frac{2}{5} = \frac{x}{20}$, $x =$

57) $\frac{2}{7} = \frac{x}{14}$, $x =$

58) $\frac{90}{6} = \frac{x}{2}$, $x =$

Similarity and Ratios

Similarity and Ratios refer to the relationship between shapes and the measurements of their corresponding parts.

Two figures are considered similar if they have the same shape, regardless of their size or orientation.

Two or more figures are considered similar if their corresponding angles are congruent and their corresponding side lengths are in proportion to each other. This means that if the ratio of the lengths of corresponding sides of two figures is always the same, the figures are similar.

Let's review an example:

The following triangles are similar. What is the value of the unknown side?

Find the corresponding sides and write a proportion. $\frac{8}{16} = \frac{6}{x}$.

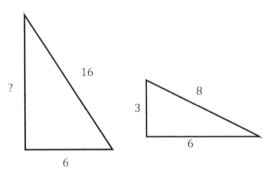

Now, use the cross-product to solve for x:

$\frac{8}{16} = \frac{6}{x} \to 8 \times x = 16 \times 6 \to 8x = 96$.

Divide both sides by 8. Then: $8x = 96 \to x = \frac{96}{8} \to x = 12$. The missing side is 12.

Another example:

Two rectangles are similar. The first is 8 meters wide and 12 meters long. The second is 4 meters wide. What is the length of the second rectangle?

To find the length of the second rectangle, we can use the concept of scale factor. Since the rectangles are similar, the ratio of their corresponding side lengths is always the same. In this case, the ratio of the length to width of the first rectangle is $12:8$ (or $3:2$). Therefore, the ratio of the length to width of the second rectangle is also $3:2$. Since the width of the second rectangle is 4 meters, we can use the proportion: $\frac{12}{8} = \frac{x}{4}$. To find the length of the rectangle x. Cross multiplying, we get: $x = (\frac{12}{8}) \times 4 = 6$ meters. Therefore, the length of the second rectangle is 6 meters.

Another example for better understanding:

Two triangles are similar. The ratio of the corresponding sides of the first triangle is $3:4:5$. The length of the longest side of the second triangle is $12 cm$. What is the length of the medium side of the second triangle?

To find the length of the medium side of the second triangle, we can use the concept of similarity and the ratio of the corresponding sides of the first triangle. Since the triangles are similar, the ratio of the corresponding sides of the second triangle is also $3:4:5$.

Since the longest side of the second triangle is $12\ cm$, we can use the proportion: $\frac{4}{5} = \frac{x}{12}$ To find the length of the medium side of the second triangle x.

Cross multiplying, we get: $x = \left(\frac{4}{5}\right) \times 12 = \frac{48}{5} = 9.6\ cm$

Therefore, the length of the medium side of the second triangle is $9.6\ cm$.

✍ Solve each problem.

59) Two rectangles are similar. One is 5 meters by 8 meters. The shorter side of the second rectangle is 12.5 meters. What is the other side of the second rectangle?

60) Two rectangles are similar. The first is 6 feet wide and 42 feet long. The second is 12 feet wide. What is the length of the second rectangle?

61) Two triangles are similar. The ratio of the corresponding sides of the first triangle is $2:4:6$. The length of the longest side of the second triangle is $18\ cm$. What is the length of the medium side of the second triangle?

62) Two triangles are similar. The ratio of the corresponding sides of the first triangle is $3:4:5$. The length of the shortest side of the second triangle is $15\ cm$. What is the length of the longest side of the second triangle?

Percent Problems

Percent problems involve using the concept of percent, which is a way of expressing a number as a fraction of 100. In percent problems, you may need to use the following concepts:

- Percentage: A number expressed as a fraction of 100, represented by the symbol "%". For example, 50% represents $\frac{50}{100}$ or 0.5.

- Percent of: The concept of finding what percentage one number is of another. For example, "50% of 80" means "$\frac{50}{100}$ of 80" or "0.5×80".

- Increase/Decrease by a certain percentage: The concept of finding the change in value when a number is increased or decreased by a certain percentage. For example, if the price of a product is increased by 10%, the new price is 110% of the original price.

- Finding the original amount: The concept of finding the original amount before a percentage increase or decrease. For example, if the price of a product is increased by 10% from $100, the new price is $110. To find the original price, you need to divide the new price by 1.1 (110% in decimal form).

Let's review some examples:

What is 50% of 80?

50% of 80 can be found by multiplying $\frac{50}{100}$ by 80. So, 50% of 80 = $0.5 \times 80 = 40$

If a product is on sale for 20% off, and the original price is $100, how much will it cost?

To find the sale price, we need to subtract 20% of the original price from the original price. 20% of $100 = $\frac{20}{100} \times \$100 = \$20$. So, the sale price = $100 − $20 = $80

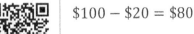

If you invest $1000 and earn a 5% interest rate, how much money will you have after one year?

To find the amount of money after one year, we need to add 5% of the initial investment to the initial investment. 5% of $1000 = $\frac{5}{100} \times \$1000 = \$50$. So, the total amount of money after one year = $1000 + $50 = $1050

If a test score is 80% and you need to get at least 90% to pass, what percent must you score on the remaining questions to pass?

To find the percentage you need to score on the remaining questions to pass, we need to subtract the current percentage score from the passing percentage. 90% − 80% = 10%. So, you need to score at least 10% on the remaining questions to pass the test.

What is 30% of 90?

In this problem, we have the percent (30%) and the base (90) and we are looking for the "part".

Use this formula: $Part = Percent \times Base$.

Then: $Part = 30\% \times 90 = \frac{30}{100} \times 90 = 0.30 \times 90 = 27$.

The answer: 30% of 90 is 27.

✒ Solve each problem.

63) What is 19% of 100?

64) 38 is what percent of 150?

65) 22 is 45% of what number?

66) What is 8% of 280?

67) 14 is 35% of what number?

68) 49 is what percent of 300?

Percent of Increase and Decrease

Percent of increase and decrease refers to the concept of finding the change in value when a number is increased or decreased by a certain percentage.

Percent of increase is calculated by finding the ratio of the increase to the original value, and then multiplying that ratio by 100 to express it as a percentage. The formula is: $\frac{(New\ value\ -\ Original\ value)}{Original\ value} \times 100 =$ Percent of increase

Percent of decrease is calculated by finding the ratio of the decrease to the original value, and then multiplying that ratio by 100 to express it as a percentage. The formula is: $\frac{(Original\ value - New\ value)}{Original\ value} \times 100 =$ Percent of decrease

For example,

- If the price of a product is increased from $100 to $120, the percent of increase can be calculated as $\frac{(120\ -\ 100)}{100} \times 100 = 20\%$

- If the price of a product is decreased from $100 to $80, the percent of decrease can be calculated as $\frac{(100-80)}{100} \times 100 = 20\%$

Let's review some examples:

The price of a product was increased from $50 to $60. What is the percentage of increase?

The percent of increase can be calculated as $\frac{(60-50)}{50} \times 100 = 20\%$. The price of the product increased by 20%.

The population of a city decreased from 100,000 to 80,000. What is the percentage of decrease?

The percent of the decrease can be calculated as $\frac{(100{,}000 - 80{,}000)}{100{,}0000} \times 100 = 20\%$. The population of the city decreased by 20%.

Worker's salary was decreased from $40,000 to $35,000. What is the percent of the decrease?

The percent of the decrease can be calculated as $\frac{(40{,}000 - 35{,}000)}{40{,}0000} \times 100 = 12.5\%$.

The worker's salary decreased by 12.5%.

The price of a table decreased from $200 to $150. What is the percent of the decrease?

Use this formula:

$Percent\ of\ change = \frac{new\ number - original\ number}{original\ number} \times 100$.

Then: $\frac{150-200}{200} \times 100 = \frac{-50}{200} \times 100 = -25$.

The percentage decrease is 25. (The negative sign means a percentage decrease.)

Therefore, the price of the table decreased by 25%.

✎ Solve each problem.

69) Find 720 decreased by 20%.
70) James puts $1,800 in a bank account. He takes it out 7 years later. It's now worth $2430. Find the percentage increase.
71) Find 810 increased by 15%.
72) A chemical solution contains 37% alcohol. If there is 111 ml of alcohol, what is the volume of the solution?

Discount, Tax, and Tip

Discount:

To solve Discount problems:

1. Express the discount percentage as a decimal by dividing it by 100.

2. Multiply the decimal value by the original price to find the amount of the discount.

3. Subtract the discount amount from the original price to find the final or selling price.

Let's review some examples of discount problems:

If an item is priced at $100 and has a 20% discount, the discount amount is $(\frac{20}{100}) \times 100 = \$20$. The final price is $\$100 - \$20 = \$80$

An item is priced at $200 and has a 25% discount.

To find the discount amount, we would calculate $(\frac{25}{100}) \times \$200 = \$50$. The final price of the item would be $\$200 - \$50 = \$150$.

With a 30% discount, Ella saved $60 on a dress. What was the original price of the dress? Let x be the original price of the dress. Then: 30% of $x = 60$.

Write an equation and solve for x:

$0.30 \times x = 60 \rightarrow x = \frac{60}{0.30} = 200$. The original price of the dress was $200.

Tax:

To solve Tax problems:

1. Express the tax rate as a decimal by dividing it by 100.

2. Multiply the decimal value by the taxable amount to find the amount of tax owed.

Let's review some examples of tax problems:

Integers and Order of Operations

If the taxable amount is $1000 and the tax rate is 10%, the amount of tax owed is $(\frac{10}{100}) \times 1000 = \$100$

A person has an income of $50,000 and the tax rate is 20%.

To find the amount of tax owed, we would calculate $(\frac{20}{100}) \times \$50,000 = \$10,000$.

Tip:

To solve Tip problems:

1. Express the tip percentage as a decimal by dividing it by 100.
2. Multiply the decimal value by the final price (or selling price) to find the amount of the tip.

Let's review some examples of tax problems:

If the final price of an item is $50 and the tip rate is 15%, the amount of tip is $(\frac{15}{100}) \times 50 = \$7.50$

A restaurant bill comes out to $100 and the customer wants to leave a 20% tip.

To find the tip amount, we would calculate $(\frac{20}{100}) \times \$100 = \$20$

Nicole and her friends went out to eat at a restaurant. If their bill was $60.00 and they gave their server a 15% tip, how much did they pay altogether?

First, find the tip. To find the tip, multiply the rate by the bill amount.

Tip $= 60 \times 0.15 = 9$. The final price is: $\$60 + \$9 = \$69$.

✎ Find the selling price of each item.

73) The original price of a phone: $850. Tax: 9.5%, Selling price: $_____

74) Sara orders a $18 lunch. The tax rate is 9% and she wants to leave a 10% tip. What will be her total cost? $ _____

75) If your cell phone bill was $93.85 and there is a 6% late fee, how much will your bill be? $ _____

bit.ly/2Je5lo0

Find more at

Simple Interest

Simple interest is a method of calculating the interest on a loan or investment where the interest is based only on the original principal or the initial amount invested. The formula for calculating simple interest is:

$$\text{Simple Interest} = \frac{\text{Principal} \times \text{Interest Rate} \times \text{Time}}{100}$$

Where:

- Principal is the initial amount of the loan or investment
- Interest Rate is the percentage of the interest that is charged on the principal
- Time is the length of the loan or investment, usually measured in years

For example, if you invest $1,000 at a 5% interest rate for 1 year, the simple interest would be: $\frac{(1,000 \times 5 \times 1)}{100} = \$50$

It's important to note that simple interest does not take into account the effect of compounding, which can significantly affect the total interest earned over time.

Simple interest is commonly used for short-term loans or investments, such as personal loans, car loans, and savings accounts.

Let's review some examples:

You invest $5,000 at an interest rate of 4% for 2 years. How much interest will you earn?

The simple interest can be calculated as $\frac{(5,000 \times 4 \times 2)}{100} = \$400$. You will earn $400 in interest.

A company borrows $25,000 at an interest rate of 6% for 3 years. How much interest will they pay?

The simple interest can be calculated as $\frac{(25{,}000 \times 6 \times 3)}{100} = \$4{,}500$. The company will pay \$4,500 in interest.

You deposit \$1000 in a savings account with an interest rate of 2% per year. How much interest will you earn after 3 years?

The simple interest can be calculated as $\frac{(1000 \times 2 \times 3)}{100} = \$60$. You will earn \$60 in interest after 3 years.

Find simple interest for \$1,600 at 6% for 6 years.

Use Interest formula:

$I = prt$ ($P = \$1{,}600$, $r = 6\% = \frac{6}{100} = 0.06$ and $t = 6$).

Then: $I = 1{,}600 \times 0.06 \times 6 = \$576$.

✎ Determine the simple interest for the following loans.

76) \$1300 at 7% for 11 years. \$_____
77) \$800 at 12% for 3 years. \$_____

✎ Solve.

78) A new car, valued at \$20,000, depreciates at 8% per year. What is the value of the car one year after purchase? \$_____

79) Jake puts \$12,000 into an investment yielding 2.5% annual simple interest; he leaves the money in for seven years. How much interest does Jake get at the end of those seven years? \$ _____

bit.ly/3nJIi3D
Find more at

The Distributive Property

The distributive property is a rule in mathematics that allows you to simplify expressions by distributing a term outside of parentheses to each term inside the parentheses. The distributive property is often used to simplify expressions that include both multiplication and addition or subtraction.

The distributive property can be applied in the following ways:
- $a(b + c) = ab + ac$
- $a(b - c) = ab - ac$

Here, "a" is a constant or a variable, and "b" and "c" are also variables or constants. The distributive property states that you can multiply "a" by "b" and "c" individually and then add or subtract them to get the same result as multiplying "a" by the expression inside the parentheses.

For example, consider the expression $4(3 + 2)$. If we use the distributive property, we can simplify it to $4(3) + 4(2) = 12 + 8 = 20$. Instead of multiplying the entire expression inside the parentheses, we can split it into two separate terms and multiply each term individually.

Let's review this concept with an example:

Simplify the expression $3x(2y + 4)$

Using the distributive property, we can simplify the expression by distributing the $3x$ to each term inside the parentheses:

$$3x(2y + 4) = 3x(2y) + 3x(4) = 6xy + 12x$$

So, the simplified expression is $6xy + 12x$.

The distributive property is a fundamental concept in algebra and is used in many different areas of mathematics, including factoring polynomials, solving equations, and simplifying complex expressions.

Integers and Order of Operations

Keep in mind that the distributive property is a property of arithmetic, and it applies to both real and complex numbers. It is a fundamental property that is widely used in mathematical operations, and it is a key concept to understand when working with algebraic expressions.

Another example: Simply. $(6)(4x - 5) - 8x$

First, simplify $(6)(4x - 5)$ using the distributive property.

Then: $(6)(4x - 5) = 24x - 30$.

Now combine like terms: $(6)(4x - 5) - 8x = 24x - 30 - 8x$.

In this expression, $24x$ and $-8x$ are "like" terms and we can combine them. $24x - 8x = 16x$. Then: $24x - 30 - 8x = 16x - 30$.

One more example:

Simply. $(-3)(3x - 5) + 4x$

First, simplify $(-3)(3x - 5)$ using distributive property.

Then: $(-3)(3x - 5) = -9x + 15$.

Now combine like terms: $(-3)(3x - 5) + 4x = -9x + 15 + 4x$.

In this expression, $-9x$ and $4x$ are "like" terms and we can combine them. $-9x + 4x = -5x$. Then: $-9x + 15 + 4x = -5x + 15$.

✎ Use the distributive property to simplify each expression.

80) $4(x + 3) =$ _____

81) $-2(3x - 1) =$ _____

82) $(4 - x)2 =$ _____

83) $(-4)(2x + 1) =$ _____

84) $(-4x - 8)(-2) =$ _____

85) $-(-7 + 5x) =$ _____

Approximating Irrational Numbers

Irrational numbers are numbers that cannot be expressed as a fraction, and they are characterized by being non-repeating, non-terminating decimals. They do not have a specific point on the number line. Examples of irrational numbers include the square roots of non-perfect square numbers.

To locate irrational numbers on a number line, we use approximations of these numbers. These approximations allow us to place the irrational numbers in a specific location on the number line, even though they do not have a precise point.

Question: Are all square roots irrational numbers?

- No, only the square roots of numbers that are not perfect squares are irrational numbers. The square roots of perfect squares (like $4, 9, 16$, etc.) are rational numbers.

Let's review an example:

Find the approximation of $\sqrt{14}$.

The square root of 14 is an irrational number and cannot be expressed as a fraction. However, we can approximate its value by finding the two closest perfect squares to 14 and taking the average of their square roots (The closest perfect squares to 14 are 9 and 16, and the square root of 9 is 3 and the square root of 16 is 4. Since 14 is closer to 16 than 9, the approximation of $\sqrt{14}$ is between 3 and 4, so we can use 3.5 as the first approximation. But since 14 is closer to 16 than 9, a better approximation is 3.7). Then: $\sqrt{14} \approx 3.7$.

Another examples:

Find the approximation of $\sqrt{32}$.

The square root of 32 is an irrational number and cannot be expressed as a fraction. However, we can approximate its value by finding the two closest perfect squares to 32 and taking the average of their square roots.

The closest perfect squares to 32 are 25 and 36, and the square root of 25 is 5 and the square root of 36 is 6. Since 32 is closer to 36 than 25, the approximation of $\sqrt{32}$ is between 5 and 6, so a better approximation is 5.7. Then: $\sqrt{32} \approx 5.7$.

Find the approximation of $\sqrt{7}$.

Since 2 is not a perfect square, $\sqrt{7}$ is irrational. To approximate $\sqrt{7}$, first, we need to find the two consecutive perfect squares that 7 is between. We can do this by writing this inequality: $4 < 7 < 9$. Now take the square root of each number: $\sqrt{4} < \sqrt{7} < \sqrt{9}$. Simplify the square roots of perfect squares:

$2 < \sqrt{7} < 3$, then $\sqrt{7}$ is between 2 and 3.

To find a better estimate, choose some numbers between 2 and 3. Let's choose 2.5, 2.6, and 2.7 → $2.5^2 = 6.25$, $2.6^2 = 6.76$, $2.7^2 = 7.29$, 6.76 is closer to 7. Then: $\sqrt{7} \approx 2.6$.

✎ Find the approximation of each.

86) $\sqrt{48} \approx$ ____

87) $\sqrt{79} \approx$ ____

88) $\sqrt{5} \approx$ ____

89) $\sqrt{125} \approx$ ____

✎ Find the approximation of each and locate them approximately on a number line diagram.

90) $\sqrt{70} \approx$ ____

91) $\sqrt{30} \approx$ ____

Chapter 1: Answers

1) 5
2) 8
3) 2
4) −2
5) −10
6) 5
7) 0
8) 17
9) −37
10) −19
11) 2
12) 50
13) 24
14) 7
15) −10
16) −30
17) −7
18) −8
19) 3
20) 8
21) $-\frac{3}{a} = 18$
22) $10 - x = 9$
23) 4^3
24) $b^2 = 25$
25) $-5 + x = 7$
26) $(x + 1) \times 5 = 25$
27) 0
28) −4
29) −1
30) 3
31) 3
32) 6
33) −10
34) 2
35) 2
36) −5
37) −5
38) 2
39) 4
40) 10
41) −7
42) 8
43) 10
44) $\frac{4}{3}$
45) $\frac{1}{3}$
46) 14
47) 9
48) 56
49) 21
50) 5
51) 12
52) 10
53) 15

54) 9
55) 8
56) 8
57) 4
58) 30
59) 20
60) 84
61) 12
62) 25
63) 19
64) 25.3
65) 48.9
66) 22.4
67) 40
68) 16.3
69) 576
70) 35%
71) 931.5

72) 300 ml
73) $930.75
74) $21.42
75) $99.48
76) $1,001
77) $288
78) 18,400
79) $2,100
80) $4x + 12$
81) $-6x + 2$
82) $-2x + 8$
83) $-8x - 4$
84) $8x + 16$
85) $-5x + 7$
86) 6.9
87) 8.9
88) 2.2
89) 11.2

90) 8.4

91) 5.5

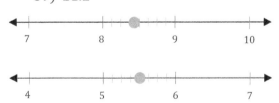

CHAPTER 2
Exponents and Variables

Math topics that you'll learn in this chapter:

- ☑ Multiplication Property of Exponents
- ☑ Division Property of Exponents
- ☑ Powers of Products and Quotients
- ☑ Zero and Negative Exponents
- ☑ Negative Exponents and Negative Bases
- ☑ Match Exponential Functions and Graphs
- ☑ Write Exponential Functions: Word Problems
- ☑ Scientific Notation
- ☑ Addition and Subtraction in Scientific Notation
- ☑ Multiplication and Division in Scientific Notation

Multiplication Property of Exponents

The Multiplication Property of Exponents is a mathematical rule that applies to expressions with the same base raised to different exponents. When you multiply two such expressions, you can simplify the product by adding the exponents. The property can be stated as follows:

For any real numbers a, m, and n: $a^m a^n = a^{m+n}$.

Here, "a" is the base, and "m" and "n" are the exponents. This property holds for both positive and negative exponents, as well as for fractions.

Here are some examples to illustrate the Multiplication Property of Exponents:

Evaluate. $(2^3)(2^4) =$

Use the Multiplication Property of Exponents. We can add the exponents:

$(2^3)(2^4) = 2^{3+4} = 2^7 = 128$.

Evaluate. $5^2 \times 5^{-3} =$

Again, add the exponents: $5^2 \times 5^{-3} = 5^{2+(-3)} = 5^{-1} = \frac{1}{5}$.

Simplify. $3^{\frac{1}{2}} \times 3^{\frac{1}{3}} =$

Add the exponents: $3^{\frac{1}{2}} \times 3^{\frac{1}{3}} = 3^{(\frac{1}{2})+(\frac{1}{3})} = 3^{\frac{5}{6}}$.

The Multiplication Property of Exponents is useful in simplifying expressions involving exponential terms with the same base, making it easier to perform calculations and solve problems.

Let's dive deeper into the Multiplication Property of Exponents and look at more examples, including ones with variables and multiple terms.

1. Variables as bases:

When you have variables as bases, the Multiplication Property of Exponents still holds. If you have the same base (variable) with different exponents, you can add the exponents when multiplying the expressions.

Simplify. $x^2 \times x^3 =$

Exponents and Variables

According to the Multiplication Property of Exponents, we can add the exponents: $x^2 \times x^3 = x^{2+3} = x^5$.

Evaluate. $a^4 \times a^{-2} =$

Add the exponents: $a^4 \times a^{-2} = a^{4-2} = a^2$.

2. Multiple terms:

When you have expressions with multiple terms, you can apply the Multiplication Property of Exponents separately for each pair of terms with the same base.

Multiply. $2^3 x^2 \times 2^4 x^5 =$

Apply the property for the numerical bases (2) and the variable bases (x) separately: $2^3 x^2 \times 2^4 x^5 = (2^{3+4}) \times (x^{2+5}) = 2^7 \times x^7 = 128 x^7$.

Evaluate. $(3^2 a^3 b) \times (3^4 a^{-1} b^2) =$

Apply the property for each pair of terms with the same base:

$(3^2 a^3 b) \times (3^4 a^{-1} b^2) = (3^{2+4}) \times \left(a^{3+(-1)}\right) \times (b^{1+2}) = 3^6 \times a^2 \times b^3 = 729 a^2 b^3$.

The Multiplication Property of Exponents helps simplify expressions and solve problems in various fields of mathematics, including algebra, calculus, and number theory. Remember that this property only applies when the bases are the same.

✎ Find the products.

1) $x^2 \times 4xy^2 =$

2) $3x^2 y \times 5x^3 y^2 =$

3) $6x^4 y^2 \times x^2 y^3 =$

4) $7xy^3 \times 2x^2 y =$

5) $-5x^5 y^5 \times x^3 y^2 =$

6) $-8x^3 y^2 \times 3x^3 y^2 =$

7) $-6x^2 y^6 \times 5x^4 y^2 =$

8) $-3x^3 y^3 \times 2x^3 y^2 =$

9) $-6x^5 y^3 \times 4x^4 y^3 =$

10) $-2x^4 y^3 \times 5x^6 y^2 =$

11) $-7y^6 \times 3x^6 y^3 =$

12) $-9x^4 \times 2x^4 y^2 =$

Division Property of Exponents

The Division Property of Exponents is a mathematical rule that applies to expressions with the same base raised to different exponents when they are being divided. When you divide two such expressions, you can simplify the quotient by subtracting the exponent of the denominator from the exponent of the numerator. The property can be stated as follows:

For any real numbers a, m, and n, and $a \neq 0$:

$$\frac{a^m}{a^n} = a^{m-n}$$

Here, "a" is the base, and "m" and "n" are the exponents. This property holds for both positive and negative exponents, as well as for fractions.

Here are some examples to illustrate the Division Property of Exponents:

Simplify. $\frac{2^5}{2^2} =$

According to the Division Property of Exponents, we can subtract the exponents:

$\frac{2^5}{2^2} = 2^{5-2} = 2^3 = 8.$

Evaluate. $\frac{7^3}{7^{-2}} =$

Subtract the exponents:

$\frac{7^3}{7^{-2}} = 7^{3-(-2)} = 7^{3+2} = 7^5 = 16{,}807.$

Simplify. $\frac{x^6}{x^3} =$

Subtract the exponents: $x^{6-3} = x^3$.

Simplify. $\frac{a^5 b^3}{a^2 b} =$

Subtract the exponents for each pair of terms with the same base:

$(a^{5-2})(b^{3-1}) = a^3 b^2.$

The Division Property of Exponents is useful in simplifying expressions involving exponential terms with the same base when dividing, making it easier to perform calculations and solve problems in various areas of mathematics, including algebra and calculus.

In summary, for the division of the exponents use following formulas:

- $\frac{x^a}{x^b} = x^{a-b}$ $(x \neq 0)$
- $\frac{x^a}{x^b} = \frac{1}{x^{b-a}}$, $(x \neq 0)$
- $\frac{1}{x^b} = x^{-b}$, $(x \neq 0)$

More examples:

Simplify. $\frac{16x^3y}{2xy^2} =$

First, cancel the common factor: $2 \to \frac{16x^3y}{2xy^2} = \frac{8x^3y}{xy^2}$.

Use Exponent's rules: $\frac{x^a}{x^b} = x^{a-b} \to \frac{x^3}{x} = x^{3-1} = x^2$ and $\frac{x^a}{x^b} = \frac{1}{x^{b-a}} \to \frac{y}{y^2} = \frac{1}{y^{2-1}} = \frac{1}{y}$.

Then: $\frac{16x^3y}{2xy^2} = \frac{8x^2}{y}$.

Simplify. $\frac{24x^8}{3x^6} =$

Use Exponent's rules: $\frac{x^a}{x^b} = x^{a-b} \to \frac{x^8}{x^6} = x^{8-6} = x^2$.

Then: $\frac{24x^8}{3x^6} = 8x^2$.

✎ Simplify.

13) $\frac{5^3 \times 5^4}{5^9 \times 5} =$

14) $\frac{3^3 \times 3^2}{7^2 \times 7} =$

15) $\frac{15x^5}{5x^3} =$

16) $\frac{16x^3}{4x^5} =$

17) $\frac{72y^2}{8x^3y^6} =$

18) $\frac{10x^3y^4}{50x^2y^3} =$

19) $\frac{13y^2}{52x^4y^4} =$

20) $\frac{50xy^3}{200x^3y^4} =$

21) $\frac{48x^2}{56x^2y^2} =$

22) $\frac{81y^6x}{54x^4y^3} =$

Powers of Products and Quotients

Powers of Products and Quotients are properties that describe how to deal with exponential expressions that involve products (multiplication) or quotients (division) raised to a power. These properties allow you to simplify expressions and perform calculations more easily. We'll discuss each property separately.

1. Power of a Product:

The Power of a Product rule states that when a product is raised to an exponent, you can distribute the exponent to each factor in the product. Mathematically, this can be expressed as:

$$(ab)^n = a^n b^n$$

Here, "a" and "b" are the factors, and "n" is the exponent.

Simplify. $(2 \times 3)^4 =$

According to the Power of a Product rule, distribute the exponent to each factor:

$$(2 \times 3)^4 = 2^4 \times 3^4 = 16 \times 81 = 1296$$

Evaluate. $(xy)^3 =$

Distribute the exponent to each factor: $(xy)^3 = x^3 y^3$.

2. Power of a Quotient:

The Power of a Quotient rule states that when a quotient is raised to an exponent, you can distribute the exponent to both the numerator and the denominator. Mathematically, this can be expressed as:

$$\left(\frac{a}{b}\right)^n = \frac{a^n}{b^n}, \text{ where } b \neq 0$$

Here, "a" is the numerator, "b" is the denominator, and "n" is the exponent.

Simplify. $\left(\frac{4}{2}\right)^3 =$

According to the Power of a Quotient rule, distribute the exponent to the numerator and the denominator: $\left(\frac{4}{3}\right)^3 = \frac{4^3}{3^3} = \frac{64}{8} = 8$.

Evaluate. $\left(\frac{x}{y}\right)^5 =$

Exponents and Variables

Distribute the exponent to the numerator and the denominator: $\frac{x^5}{y^5}$.

These properties are particularly useful when simplifying expressions or solving problems in algebra, calculus, and other areas of mathematics that involve products or quotients raised to powers. In summary:

- For any nonzero numbers a and b and any integer x,

$$(ab)^x = a^x \times b^x \text{ and } \left(\frac{a}{b}\right)^x = \frac{a^x}{b^x}$$

Let's review an example:

Simplify. $(3x^3y^2)^2 =$

Use Exponents rules: $(x^a)^b = x^{a \times b}$.

$(3x^3y^2)^2 = (3)^2(x^3)^2(y^2)^2 = 9x^{3 \times 2}y^{2 \times 2} = 9x^6y^4$.

Another example:

Simplify. $\left(\frac{2x^3}{3x^2}\right)^2 =$

First, cancel the common factor:

$x \to \left(\frac{2x^3}{3x^2}\right) = \left(\frac{2x}{3}\right)^2$.

Use Exponents rules: $\left(\frac{a}{b}\right)^c = \frac{a^c}{b^c}$.

Then: $\left(\frac{2x}{3}\right)^2 = \frac{(2x)^2}{(3)^2} = \frac{4x^2}{9}$.

✎ Solve.

23) $(x^3y^3)^2 =$

24) $(3x^3y^4)^3 =$

25) $(4x \times 6xy^3)^2 =$

26) $(5x \times 2y^3)^3 =$

27) $\left(\frac{9x}{x^3}\right)^2 =$

28) $\left(\frac{3y}{18y^2}\right)^2 =$

29) $\left(\frac{3x^2y^3}{24x^4y^2}\right)^3 =$

30) $\left(\frac{26x^5y^3}{52x^3y^5}\right)^2 =$

31) $\left(\frac{18x^7y^4}{72x^5y^2}\right)^2 =$

32) $\left(\frac{12x^6y^4}{48x^5y^3}\right)^2 =$

EffortlessMath.com

Zero and Negative Exponents

Zero and negative exponents are mathematical concepts that help simplify expressions involving the powers of numbers.

Zero exponent: The concept of a zero exponent might seem counterintuitive at first, as it's not immediately clear why any non-zero number raised to the power of zero should equal one. However, this rule is consistent with the properties of exponents and can be explained using the following rationale:

Consider the general rule for dividing numbers with the same base but different exponents:

$$\frac{a^m}{a^n} = a^{m-n}$$

Now, let's assume $m = n$: $\frac{a^m}{a^m} = a^{m-m}$.

Since $m - m = 0$, the expression simplifies to: $\frac{a^m}{a^m} = a^0$.

On the left side of the equation, any non-zero number divided by itself is equal to 1: $1 = a^0$.

This is why any non-zero number raised to the power of zero is equal to one.

Negative exponent: Negative exponents indicate that the base should be inverted (reciprocal) and raised to the positive value of the exponent. This can also be understood using the properties of exponents.

Consider the general rule for dividing numbers with the same base but different exponents, as mentioned earlier: $\frac{a^m}{a^n} = a^{m-n}$.

Now, let's assume $m = 0$: $\frac{a^0}{a^n} = a^{0-n}$.

Since $0 - n = -n$, the expression simplifies to: $\frac{a^0}{a^n} = a^{-n}$.

We already know that $a^0 = 1$ for any non-zero number a, so the expression becomes: $\frac{1}{a^n} = a^{-n}$.

Exponents and Variables

This is why a number with a negative exponent is equal to one divided by the same number with a positive exponent.

- Zero-Exponent Rule: $a^0 = 1$, this means that anything raised to the zero power is 1. For example: $(5xy)^0 = 1$. (Number zero is an exception: $0^0 = 0$.)
- A negative exponent simply means that the base is on the wrong side of the fraction line, so you need to flip the base to the other side. For instance, "x^{-2}" (pronounced as "ecks to the minus two") just means "x^2" but underneath, as in $\frac{1}{x^2}$.

More examples:

Evaluate. $\left(\frac{4}{5}\right)^{-2} =$

Use negative exponents rule:

$\left(\frac{x^a}{x^b}\right)^{-2} = \left(\frac{x^b}{x^a}\right)^2 \rightarrow \left(\frac{4}{5}\right)^{-2} = \left(\frac{5}{4}\right)^2$.

Then: $\left(\frac{5}{4}\right)^2 = \frac{5^2}{4^2} = \frac{25}{16}$.

Evaluate. $\left(\frac{3}{500}\right)^0 =$

Use zero-exponent Rule: $a^0 = 1$.

Then: $\left(\frac{3}{500}\right)^0 = 1$.

✎ Evaluate each expression.

33) $\left(\frac{1}{4}\right)^{-2} =$

34) $\left(\frac{1}{3}\right)^{-2} =$

35) $\left(\frac{1}{7}\right)^{-3} =$

36) $\left(\frac{2}{5}\right)^{-3} =$

37) $\left(\frac{2}{3}\right)^{-3} =$

38) $\left(\frac{3}{5}\right)^{-4} =$

Negative Exponents and Negative Bases

Negative exponents and negative bases are distinct concepts in mathematics, but both are essential in understanding how numbers and equations work. Let's discuss each concept in detail.

1. Negative Exponents: A negative exponent is an exponent that has a negative value. In general, if you have a number 'a' raised to the power of a negative exponent 'n', it is written as a^{-n}.

According to the rules of exponents, a number raised to a negative exponent is equivalent to the reciprocal of the number raised to the positive exponent. In other words:

$$a^{-n} = \frac{1}{a^n}$$

For example: $2^{-3} = \frac{1}{2^3} = \frac{1}{8} = 0.125$.

Negative exponents are helpful when dealing with very small numbers, especially in scientific notation. For instance, in physics or chemistry, you might encounter values such as 3×10^{-9}, which represents 3 billionths (0.000000003).

2. Negative Bases: A negative base is a number less than zero that is being raised to a certain exponent. When working with negative bases, it is essential to pay attention to the exponent's value, as it will dictate the result's sign.

Consider the base 'a' to be a negative number:

- If the exponent is an even number ($n = 2k$, where k is an integer), the result will be positive: $(-a)^n = (-a)^{2k} = (a^2)^k = a^{2k} > 0$

For example: $(-2)^4 = (2^2)^2 = 16$.

- If the exponent is an odd number ($n = 2k + 1$, where k is an integer), the result will be negative:

$$(-a)^n = (-a)^{2k+1} = -a \times (-a)^{2k} = -a \times a^{2k} = -a^{2k+1} < 0$$

EffortlessMath.com

Exponents and Variables

For example: $(-2)^3 = -2^3 = -8$.

It is important to note that the rules for negative bases and exponents can be combined.

For example, if you have a negative base raised to a negative exponent:

$$(-a)^{-n} = \frac{1}{(-a)^n}$$

In this case, you'd apply the rules for negative bases as previously explained and then take the reciprocal as directed by the negative exponent. Do the following:

- A negative exponent is the reciprocal of that number with a positive exponent. $(3)^{-2} = \frac{1}{3^2}$.
- To simplify a negative exponent, make the power positive!
- The parenthesis is important! -5^{-2} is not the same as $(-5)^{-2}$:

$$-5^{-2} = -\frac{1}{5^2} \text{ and } (-5)^{-2} = +\frac{1}{5^2}$$

Simplify. $\left(\frac{2a}{3c}\right)^{-2} =$

Use negative exponents rule: $\left(\frac{x^a}{x^b}\right)^{-2} = \left(\frac{x^b}{x^a}\right)^2 \rightarrow \left(\frac{2a}{3c}\right)^{-2} = \left(\frac{3c}{2a}\right)^2$.

Now use exponents rule: $\left(\frac{a}{b}\right)^c = \frac{a^c}{b^c} \rightarrow \left(\frac{3c}{2a}\right)^2 = \frac{3^2 c^2}{2^2 a^2}$. Then: $\frac{3^2 c^2}{2^2 a^2} = \frac{9c^2}{4a^2}$.

✎ Simplify.

39) $x^{-7} =$

40) $3y^{-5} =$

41) $15y^{-3} =$

42) $-20x^{-4} =$

43) $12a^{-3}b^5 =$

44) $25a^3 b^{-4} c^{-3} =$

45) $-4x^5 y^{-3} z^{-6} =$

46) $\frac{18y}{x^3 y^{-2}} =$

47) $\frac{20a^{-2}b}{-12c^{-4}} =$

Match Exponential Functions and Graphs

Exponential functions are mathematical functions that involve a base raised to a variable exponent. They can be represented by the general equation:
$$f(x) = a(b)^x$$
Here, 'a' is the initial value (y-intercept), 'b' is the base, and 'x' is the exponent. The base 'b' must be a positive number different from 1 to be considered an exponential function.

To match exponential functions and their graphs, follow these steps:

1. Identify the initial value (y-intercept): Find the point where the graph intersects the y-axis. This point corresponds to the initial value 'a' in the equation $f(x) = a(b)^x$.

2. Determine if the function represents growth or decay: Examine the graph to see if it increases or decreases as x increases. If the graph increases, the function represents exponential growth, and $b > 1$. If the graph decreases, the function represents exponential decay, and $0 < b < 1$.

3. Estimate the base 'b': To estimate the base 'b', find a point on the graph where x has increased by 1 (e.g., from $x = 0$ to $x = 1$). Then, calculate the ratio between the function's values at these two points. This ratio should be equal to 'b' in the equation $f(x) = a(b)^x$.

4. Write the equation: Using the information gathered from the previous steps, write the equation for the exponential function: $f(x) = a(b)^x$.

5. Match the function to the graph: Compare the equation you've derived with the given graph. Ensure the initial value, growth or decay behavior, and the horizontal asymptote match the graph. If they do, then you've successfully matched the exponential function to its graph.

Let's review an example:

Match each exponential function to its graph.

Exponents and Variables

$f(x) = \left(\frac{1}{2}\right)^x$, $g(x) = \left(\frac{3}{2}\right)^x$, $h(x) = \left(\frac{1}{3}\right)^x$

A
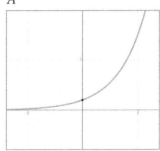

B
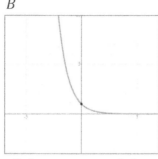

C
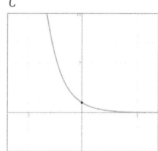

According to the base value of the exponential equation, notice that one of the functions is growing and the other two functions are decaying. So, the function $g(x) = \left(\frac{3}{2}\right)^x$ can be equivalent to graph A. Now, evaluate the two remaining functions at a few inputs to find some points on the graph. Choose a point such as -1 to plug into the equation. Therefore, start with $f(x) = \left(\frac{1}{2}\right)^x$.

$$f(-1) = \left(\frac{1}{2}\right)^{-1} = 2.$$

You can see that the ordered pair $(-1, 2)$ for the function $f(x) = \left(\frac{1}{2}\right)^x$ is equivalent to a point on graph C. In the same way, substitute -1 in $h(x) = \left(\frac{1}{3}\right)^x$. Therefore, $h(-1) = \left(\frac{1}{3}\right)^{-1} = 3$. That is, graph B represented the function $h(x)$.

✍ Match each equation to the correct graph.

48) $f(x) = -4(3)^x$, $f(x) = 2^x + 5$, $f(x) = 3(2)^x + 3$

A
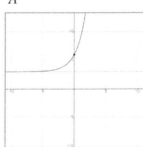

B
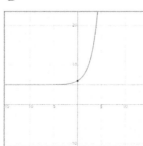

C
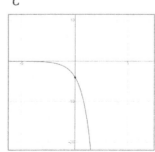

Write Exponential Functions: Word Problems

Exponential functions are mathematical expressions that model the behavior of quantities that grow or decay at a constant rate. They are particularly useful for modeling real-world phenomena such as population growth, radioactive decay, and compound interest. To solve word problems involving exponential functions, it's important to understand the basic form of an exponential function, how to set up an equation, and then solve for the unknown variable.

The general form of an exponential function is: $y = ab^x$, where:

y is the final amount after some time or number of instances.

a is the initial amount (when $x = 0$).

b is the base or growth factor ($b > 1$ for growth and $0 < b < 1$ for decay).

x is the independent variable, which represents time or number of instances.

Let's go through the process of solving an exponential word problem step by step.

Step 1: Identify the given information: Carefully read the problem and identify the information provided. Pay attention to the initial amount, the growth or decay factor, and the time or instances involved.

Step 2: Determine the form of the exponential function: Determine if the problem involves exponential growth or decay. If it's growth, $b > 1$; if it's decay, $0 < b < 1$.

Step 3: Set up the exponential function: Using the general form $y = ab^x$, plug in the given values and set up the exponential function to model the situation.

Step 4: Solve the equation: Solve the equation for the unknown variable(s), either by isolating the variable or by using logarithms.

Let's work through an example to illustrate these steps:

A bacteria culture starts with 50 bacteria and doubles in number every 3 hours. How many bacteria will be present after 12 hours?

Step 1: Identify the given information:

Initial amount (a) = 50 bacteria, Growth factor (b) = 2 (doubles every 3 hours), Time period (x) = 12 hours

Step 2: Determine the form of the exponential function. Since the bacteria culture is growing, we have an exponential growth function.

Step 3: Set up the exponential function The general form of the exponential function is $y = ab^x$. In this case, we have: $y = 50(2)^{\frac{x}{3}}$.

Step 4: Solve the equation. We want to find the number of bacteria after 12 hours, so we need to solve for y when $x = 12$:

$y = 50(2)^{\frac{12}{3}} \rightarrow y = 50(2)^4 \rightarrow y = 50(16) \rightarrow y = 800$.

After 12 hours, there will be 800 bacteria in the culture.

Let's another example:

In a laboratory sample, if it starts with 100 bacteria which can double every hour, how many bacteria will there be after 6 hours?

Since it grows at a constant ratio of 2, you have to use the formula of exponential to find the number of bacteria in this sample. Use the formula, $y = ab^n$, where $a = 100$ is the initial number of bacteria, and b is equal to 2. Substitute the given value in the formula, then: $y = 100(2)^n$.

Now, plug $n = 6$ into the equation. Therefore, $y = 100(2)^6 = 100(64) = 6,400$.

✏ Write each number in scientific notation.

49) As of 2019, the world population is 8.716 billion and growing at a rate of 1.2% per year. Write an equation to model population growth, where $p(t)$ is the population in billions of people and t is the time in years.

50) You decide to buy a used car that costs $20,000. You have heard that the car may depreciate at a rate of 10% per year. At this rate, how much will the car be worth in 6 years?

Scientific Notation

In scientific notation, a number is represented as a decimal multiplied by a power of 10. The decimal, also known as the coefficient, is a number between 1 and 10, and the exponent tells us how many times the decimal must be multiplied by 10 to obtain the original number.

For example, the number 0.00005 can be written in scientific notation as 5×10^{-5}. To convert this to standard form, we move the decimal point 5 places to the left, giving us 0.00005. Similarly, a number like 35,000 can be written in scientific notation as 3.5×10^4. To convert this to standard form, we move the decimal point 4 places to the right, giving us 35,000.

It's important to note that in scientific notation, the decimal point is always placed after the first non-zero digit. This helps to make the numbers more readable and easier to compare. This notation is particularly useful in scientific and technical fields, where large or small numbers are commonly encountered.

Here are a few examples of numbers written in scientific notation and their equivalent in standard form:

$3.2 \times 10^3 = 3200$

$4.5 \times 10^{-7} = 0.00000045$

$6.78 \times 10^5 = 678000$

$2 \times 10^{-9} = 0.000000002$

$9 \times 10^6 = 9000000$

In the first example, the coefficient is 3.2 and the exponent is 3, so we move the decimal point 3 places to the right to get 3200.

In the second example, the coefficient is 4.5 and the exponent is -7, so we move the decimal point 7 places to the left to get 0.00000045.

Similarly, the third example, 6.78×10^5 is 678000, and so on.

It's worth noting that the exponent tells us how many times we have to move the decimal point, if the exponent is positive, we move the

decimal point to the right and if the exponent is negative, we move the decimal point to the left.

Let's understand this more with another example:

Write 0.0034 in scientific notation.

First, move the decimal point to the right so you have a number between 1 and 10. That number is 3.4.

Now, determine how many places the decimal moved in step 1 by the power of 10. We moved the decimal point 3 digits to the right. Then: 10^{-3} → When the decimal moved to the right, the exponent is negative. Then: $0.0034 = 3.4 \times 10^{-3}$.

Another example:

Write 4.8×10^{-4} in standard notation.

The exponent is negative 4. Then, move the decimal point to the left four digits. (Remember $4.8 = 00004.8$.) When the decimal moved to the right, the exponent is negative. Then: $4.8 \times 10^{-4} = 0.00048$.

One more example:

Write 0.00024 in scientific notation.

First, move the decimal point to the right so you have a number between 1 and 10. That number is 2.4. Now, determine how many places the decimal moved in step 1 by the power of 10. We moved the decimal point 4 digits to the right. Then: 10^{-4} → When the decimal moved to the right, the exponent is negative. Then: $0.00024 = 2.4 \times 10^{-4}$.

✎ Write each number in scientific notation.

51) $0.00412 =$

52) $0.000053 =$

53) $66,000 =$

54) $72,000,000 =$

Addition and Subtraction in Scientific Notation

In order to add or subtract numbers written in scientific notation, they must have the same exponent. This is because the exponent represents the power of 10, and adding or subtracting numbers with different exponents would not make sense mathematically.

For example, to add 2.5×10^4 and 3.6×10^4, we first align the exponents, which in this case are both 4. Then we add the coefficients, $2.5 + 3.6 = 6.1$.

Finally, we write the result in scientific notation: 6.1×10^4.

To subtract, the process is similar. We align the exponents and subtract the coefficients.

For example, to subtract 6.1×10^4 and 3.6×10^4, we align the exponents which are both 4, then we subtract the coefficients $6.1 - 3.6 = 2.5$ and write the result in scientific notation 2.5×10^4.

It's important to keep in mind that the coefficient, m, should always be a number between 1 and 10, and if it's not, you'll need to adjust the coefficient and the exponent accordingly.

Let's review another example:

Simplify. $(3.4 \times 10^4) + (2.6 \times 10^5) =$

3.4×10^4 and 2.6×10^5 have different exponents, so we cannot add or subtract them directly.

To add them, we have to rewrite them so that they have the same exponent, for example we can rewrite 3.4×10^4 as 0.34×10^5 so that the exponents match and we can add the coefficient, $0.34 + 2.6 = 2.94$. So, the result would be

$$2.94 \times 10^5$$

Let's another example:

Simplify. $(5.4 \times 10^4) - (2.6 \times 10^3) =$

Exponents and Variables

To subtract 5.4×10^4 and 2.6×10^3, we have to rewrite them so that they have the same exponent.

In this case, we can rewrite 2.6×10^3 as 0.26×10^4, so that the exponents match and we can subtract the coefficient, $5.4 - 0.26 = 5.14$. So, the result would be
$$5.14 \times 10^4$$

More examples:

Simplify. $(0.34 \times 10^2) + (46 \times 10^{-1}) =$

First, write each term to the same exponent:

$0.34 \times 10^2 = 3.4 \times 10$

$4.6 \times 10^{-1} = 0.46 \times 10$.

Next, add the coefficient: $3.4 + 0.46 = 3.86$.

Now, we get: $(0.34 \times 10^2) + (46 \times 10^{-1}) = 3.86 \times 10$.

Simplify. $(10^{-3}) - (4.2 \times 10^{-2}) + (3700 \times 10^{-5}) =$

First, rewrite the term 0.0037×10^{-5} as standard notation:

$3700 \times 10^{-5} = 3.7 \times 10^{-2}$

Then, write other terms to the same exponent with 10^{-2}: $10^{-3} = 0.1 \times 10^{-2}$

Next, we have:

$(0.1 \times 10^{-2}) - (4.2 \times 10^{-2}) + (3700 \times 10^{-5}) = (0.1 - 4.2 + 3.7) \times 10^{-2}$
$$= -0.4 \times 10^{-2} = -4 \times 10^{-3}$$

✎ Write the answer in scientific notation.

55) $6 \times 10^4 + 10 \times 10^4 =$ _____

56) $7.2 \times 10^6 - 3.3 \times 10^6 =$ _____

57) $2.23 \times 10^7 + 5.2 \times 10^7 =$ _____

58) $8.3 \times 10^9 - 5.6 \times 10^8 =$ _____

59) $1.4 \times 10^2 + 7.4 \times 10^5 =$ _____

60) $9.6 \times 10^6 - 3 \times 10^4 =$ _____

Multiplication and Division in Scientific Notation

Multiplication: When multiplying numbers in scientific notation, we multiply the coefficients and add the exponents.

For example, to multiply $(4.3 \times 10^3) \times (2.6 \times 10^5)$, we first multiply 4.3 and 2.6 to get 11.18, then add the exponents 3 and 5 to get 8. The result is 11.18×10^8. The coefficient, 11.18, in the example provided is greater than 10.

In scientific notation, the coefficient should always be between 1 and 10. When the coefficient is greater than 10, it's called an "overflow" and it needs to be adjusted by moving the decimal point to the left and increasing the exponent by one for each move.

In this case, we move the decimal point to the left once, which would give us 1.118×10^9. This is now in the correct form for scientific notation.

Division: When dividing numbers in scientific notation, we divide the coefficients and subtract the exponents of the denominator from the numerator.

For example, to divide $\frac{4.3 \times 10^3}{2.6 \times 10^5}$, we first divide 4.3 by 2.6 to get 1.65, then subtract the exponent of the denominator (5) from the exponent of the numerator (3) to get -2. The result is 1.65×10^{-2}.

It's important to keep in mind that the exponent tells us how many times we have to move the decimal point, if the exponent is positive, we move the decimal point to the right and if the exponent is negative, we move the decimal point to the left, this is why in multiplication we add the exponents and in division we subtract them.

When dividing numbers in scientific notation, we divide the coefficients and subtract the exponents of the denominator from the numerator. This is because the exponents represent the power of 10, and dividing by a power of 10 is the same as moving the decimal point to the left or right, depending on the sign of the exponent.

Let's another example:

Simplify. $\frac{8 \times 10^3}{2 \times 10^2} =$

In this example, we first divide 8 by 2 to get 4, then subtract the exponent of the denominator (2) from the exponent of the numerator (3) to get 1.

The result is 4×10^1, which is the same as moving the decimal point one place to the right.

$$\frac{8 \times 10^3}{2 \times 10^2} = 4 \times 10^1$$

More examples:

Simplify. $\left(\frac{6 \times 10^{-8}}{3 \times 10^{-6}}\right) =$

In this example, we first divide 6 by 3 to get 2, then subtract the exponent of the denominator (-8) from the exponent of the numerator (-6) to get -2.

The result is 2×10^{-2}, which is the same as moving the decimal point one place to the left.

$$\left(\frac{6 \times 10^{-8}}{3 \times 10^{-6}}\right) = 2 \times 10^{-2}$$

✒ Simplify. Write the answer in scientific notation.

61) $\frac{9 \times 10^8}{3 \times 10^7} = $ _____

62) $(3 \times 10^{-8})(7 \times 10^{10}) = $ _____

63) $(9 \times 10^{-3})(4.2 \times 10^6) = $ _____

64) $\frac{125 \times 10^9}{50 \times 10^{12}} = $ _____

65) $\frac{2.8 \times 10^{12}}{0.4 \times 10^{20}} = $ _____

66) $(5.6 \times 10^{12})(3 \times 10^{-7}) = $ _____

Chapter 2: Answers

1) $4x^3y^2$
2) $15x^5y^3$
3) $6x^6y^5$
4) $14x^3y^4$
5) $-5x^8y^7$
6) $-24x^6y^4$
7) $-30x^6y^8$
8) $-6x^6y^5$
9) $-24x^9y^6$
10) $-10x^{10}y^5$
11) $-21x^6y^9$
12) $-18x^8y^2$
13) $\frac{1}{125}$
14) $\frac{243}{343}$
15) $3x^2$
16) $\frac{4}{x^2}$
17) $\frac{9}{x^3y^4}$
18) $\frac{xy}{5}$
19) $\frac{1}{4x^4y^2}$
20) $\frac{1}{4x^2y}$
21) $\frac{6}{7y^2}$
22) $\frac{3y^3}{2x^3}$
23) x^6y^6
24) $27x^9y^{12}$
25) $576x^4y^6$
26) $1{,}000x^3y^9$
27) $\frac{81}{x^4}$
28) $\frac{1}{36y^2}$
29) $\frac{y^3}{512x^6}$
30) $\frac{x^4}{4y^4}$
31) $\frac{x^4y^4}{16}$
32) $\frac{x^2y^2}{16}$
33) 16
34) 9
35) 343
36) $\frac{125}{8}$
37) $\frac{27}{8}$
38) $\frac{625}{81}$
39) $\frac{1}{x^7}$
40) $\frac{3}{y^5}$
41) $\frac{15}{y^3}$
42) $-\frac{20}{x^4}$
43) $\frac{12b^5}{a^3}$
44) $\frac{25a^3}{b^4c^3}$
45) $-\frac{4x^5}{y^3z^6}$
46) $\frac{18y^3}{x^3}$
47) $-\frac{5bc^4}{3a^2}$
48) $A: f(x) = 3(2)^x + 3$,
    $B: f(x) = 2^x + 5$,
    $C: f(x) = -4(3)^x$
49) $p(t) = 8.716(1 + 0.012)^t$
50) $A = 20{,}000(1 - 0.1)^6$
51) 4.12×10^{-3}
52) 5.3×10^{-5}
53) 6.6×10^4
54) 7.2×10^7
55) 1.6×10^5
56) 3.9×10^6
57) 7.43×10^7
58) 7.74×10^9
59) 7.4014×10^5
60) 9.57×10^6
61) 3×10^1
62) 2.1×10^3
63) 3.78×10^4
64) 2.5×10^{-3}
65) 7×10^{-8}
66) 1.68×10^6

CHAPTER
3 Expressions and Equations

Math topics that you'll learn in this chapter:

- ☑ Simplifying Variable Expressions
- ☑ Evaluating One Variable
- ☑ Evaluating Two Variables
- ☑ One–Step Equations
- ☑ Multi–Step Equations
- ☑ Rearrange Multi-Variable Equations
- ☑ Finding Midpoint
- ☑ Finding the Distance of Two Points
- ☑ Graphing Absolute Value Equations

Simplifying Variable Expressions

In algebra, a variable is a symbol that represents an unknown value. Variables are often represented by letters such as x, y, z, a, b, c, m, and n. These letters can take on different values, allowing us to express a wide range of mathematical relationships.

An algebraic expression is a mathematical phrase that contains variables, integers, and mathematical operations.

For example, the expression "$3x + 2y$" is an algebraic expression because it contains variables (x and y), integers (3 and 2), and mathematical operations (+). In algebra, we can simplify expressions by combining "like" terms. Like terms are terms that have the same variable and the same exponent.

For example, the terms "$4x$" and "$2x$" are like terms because they have the same variable (x) and the same exponent (1). We can combine these terms by adding their coefficients (4 and 2) to get "$6x$". This simplifies the expression and makes it easier to work with.

Example of simplifying an expression:

Original expression: $3x^2 + 5x - 2x^2 - 8x + 6 =$

　　Step 1: Combine like terms: $3x^2 - 2x^2 = x^2$.

　　Step 2: Combine like terms: $5x - 8x = -3x$.

　　Step 3: Combine like terms: $x^2 - 3x + 6$.

Simplified expression: $x^2 - 3x + 6$.

This is a basic example of simplifying algebraic expression by combining like terms. Algebraic simplification can be more complex and can include expanding polynomials and factoring expressions.

Let's take a look at another example:

Simplify. $(-2 + 5x^2 + 4x^2 + 7x) =$

The answer is:

$(-2 + 5x^2 + 4x^2 + 7x) = (5x^2 + 4x^2 + 7x - 2) = (9x^2 + 7x - 2)$.

In this expression, I have combined the like terms that have the same variable and exponent ($5x^2$ and $4x^2$) and also combined the coefficients of the like terms of x is ($7x$) and constant term (-2) which results in the final simplified expression of $9x^2 + 7x - 2$.

Another example:

Simplify. $(4x + 2x + 4) =$

In this expression, there are three terms: $4x$, $2x$, and 4.

Two terms are "like" terms: $4x$ and $2x$.

Combine like terms. $4x + 2x = 6x$.

Then: $(4x + 2x + 4) = 6x + 4$.

(Remember you cannot combine variables and numbers.)

Let's solve another example:

Simplify. $-2x^2 - 5x + 4x^2 - 9 =$

Combine "like" terms: $-2x^2 + 4x^2 = 2x^2$.

Then: $-2x^2 - 5x + 4x^2 - 9 = 2x^2 - 5x - 9$.

✎ Simplify each expression.

1) $(3 + 4x - 1) =$

2) $(-5 - 2x + 7) =$

3) $(12x - 5x - 4) =$

4) $(-16x + 24x - 9) =$

5) $(6x + 5 - 15x) =$

6) $2 + 5x - 8x - 6 =$

7) $5x + 10 - 3x - 22 =$

8) $2x^2 - 7x - 3x^2 + 4x + 6 =$

Evaluating One Variable

Evaluating an algebraic expression with one variable involves substituting a specific value for that variable and then performing the arithmetic operations according to the order of operations.

For example, if we have an expression $2x + 3$ and we want to evaluate it for $x = 4$, we would substitute 4 in for x, getting $2(4) + 3 = 8 + 3 = 11$.

Another example is if we have an expression $5x^2 - 3x + 2$, and we want to evaluate it for $x = -2$, we would substitute -2 in for x, getting $5(-2)^2 - 3(-2) + 2 = 20 + 6 + 2 = 28$.

Keep in mind that evaluating an algebraic expression with one variable gives a specific numerical value, whereas simplifying an algebraic expression gives an algebraic expression in the simplest form.

Let's take a look at some examples:

Calculate this expression for $x = 3$: $13 + 5x$.

To calculate the expression "$13 + 5x$" when $x = 3$, we need to substitute the value of 3 for the variable x in the expression.

So, $13 + 5x = 13 + 5(3) = 13 + 15 = 28$.

Therefore, for $x = 3$, the expression evaluates to 28.

Another example:

Solve this expression for $x = -4$: $13 - 5x$.

To solve the expression "$13 - 5x$" for "$x = -4$", we need to substitute the value of -4 for the variable x in the expression.

So, $13 - 5x = 13 - 5(-4) = 13 + 20 = 33$.

Therefore, for $x = -4$, the expression evaluates to 33.

Let's do another example:

Calculate this expression for $x = 2$: $8 + 2x$.

First, substitute 2 for x.

Then: $8 + 2x = 8 + 2(2)$.

Now, use the order of operation to find the answer:

$8 + 2(2) = 8 + 4 = 12$.

Another example:

Evaluate this expression for $x = -1$: $4x - 8$.

First, substitute -1 for x.

Then: $4x - 8 = 4(-1) - 8$.

Now, use the order of operation to find the answer:

$4(-1) - 8 = -4 - 8 = -12$.

To better understand this issue, consider this example:

Find the value of this expression when $x = 4$: $(16 - 5x)$.

First, substitute 4 for x.

Then: $16 - 5x = 16 - 5(4) = 16 - 20 = -4$.

✎ Evaluate each expression using the value given.

9) $x = 2 \rightarrow x + 3 =$

10) $x = -4 \rightarrow 8 + x =$

11) $y = 0 \rightarrow 2y + 3 =$

12) $x = -2 \rightarrow -\frac{1}{2}x =$

13) $a = -7 \rightarrow 2a + 14 =$

14) $x = 6 \rightarrow 5 + 2x =$

15) $x = 1.5 \rightarrow -2x + 1 =$

16) $x = 5 \rightarrow 12 - 3x =$

Evaluating Two Variables

To evaluate an algebraic expression, we substitute a specific numerical value for each variable in the expression. Then, we perform the arithmetic operations according to the order of operations to find the numerical value of the expression. This process is also known as "plugging in" the values and solving the equation. For example, if we have an expression $2x + y + 3$ and we want to evaluate it for $x = 4$ and $y = 5$, we would substitute 4 in for x and 5 in for y, getting $2(4) + 5 + 3 = 8 + 5 + 3 = 16$.

Let's review some examples:

Calculate this expression for $a = 4$ and $b = -3$: $(5a - 6b)$.

To calculate the expression "$5a - 6b$" when $a = 4$ and $b = -3$, we need to substitute the values of 4 for the variable a and -3 for the variable b in the expression. So, $5a - 6b = 5(4) - 6(-3) = 20 + 18 = 38$.

Therefore, for $a = 4$ and $b = -3$, the expression evaluates to 38.

Evaluate this expression for $x = -5$ and $y = -3$: $(4x + 2y)$.

To evaluate the expression "$4x + 2y$" for $x = -5$ and $y = -3$, we need to substitute the values of -5 for the variable x and -3 for the variable y in the expression. So, $4x + 2y = 4(-5) + 2(-3) = -20 - 6 = -26$.

Therefore, for $x = -5$ and $y = -3$, the expression evaluates to -26.

Find the value of this expression $3(3a - 6b)$, when $a = -2$ and $b = 5$.

To find the value of the expression "$3(3a - 6b)$" when $a = -2$ and $b = 5$, we need to substitute the values of -2 for the variable a and 5 for the variable b in the expression. So, $3(3a - 6b) = 3(3(-2) - 6(5)) = 3(-6 - 30) = 3(-36) = -108$.

Therefore, for $a = -2$ and $b = 5$, the expression evaluates to -108.

Expressions and Equations

Calculate this expression for $a = 2$ and $b = -1$: $(4a - 3b)$.

First, substitute 2 for a, and -1 for b.

Then: $4a - 3b = 4(2) - 3(-1)$.

Now, use the order of operation to find the answer:

$4(2) - 3(-1) = 8 + 3 = 11$.

Evaluate this expression for $x = -2$ and $y = 2$: $(3x + 6y)$.

Substitute -2 for x, and 2 for y.

Then: $3x + 6y = 3(-2) + 6(2) = -6 + 12 = 6$.

Find the value of this expression $2(6a - 5b)$, when $a = -1$ and $b = 4$.

Substitute -1 for a, and 4 for b.

Then: $2(6a - 5b) = 2(6(-1) - 5(4)) = 2(-6 - 20) = 2(-26) = -52$.

Evaluate this expression. $-7x - 2y$, $x = 4$, $y = -3$.

Substitute 4 for x, and -3 for y and simplify.

Then: $-7x - 2y = -7(4) - 2(-3) = -28 + 6 = -22$.

✏️ Evaluate each expression using the values given.

17) $x = -1$ and $y = 3 \rightarrow 2x + y =$

18) $x = 7$ and $y = 2 \rightarrow x - 4y =$

19) $a = -3$ and $b = 3 \rightarrow a + b =$

20) $r = 2$ and $s = 2 \rightarrow -r + 2s - 4 =$

21) $x = -8$ and $y = 4 \rightarrow x - y + 2 =$

22) $x = 1$ and $y = -1 \rightarrow 3x + 4y =$

One–Step Equations

An equation is a statement of equality between two expressions. For example, in the equation "$ax = b$", the expression "ax" is equal to the expression "b". The variable "x" is the unknown value that we want to find.

Solving an equation means determining the value of the variable that makes the equation true. To solve a one-step equation, we need to use the inverse operation of the one being performed on the variable. Inverse operations are operations that undo each other, for example, addition and subtraction, multiplication, and division.

The inverse operations are:
- Addition and subtraction
- Multiplication and division

For example, if we have an equation $x + 3 = 7$, we can solve for x by subtracting 3 from both sides: $x + 3 = 7$.

Then $x + 3 - 3 = 7 - 3 = 4$.

Another example:

If we have an equation $\frac{x}{5} = 6$, we can solve for x by multiplying both sides by 5: $\frac{x}{5} = 6 \rightarrow x = 30$.

It is important to note that solving an equation means finding the value of the variable that makes the equation true.

Let's review another example:

Solve this equation for x: $3x = 10.5 \rightarrow x = ?$

To solve the equation "$3x = 10.5$" for x, we need to isolate x on one side of the equation. $3x = 10.5$

To do this, we divide both sides of the equation by 3: $x = \frac{10.5}{3} \rightarrow x = 3.5$

Therefore, $x = 3.5$ is the solution of the equation $3x = 10.5$.

Another example:

Solve this equation for x: $4x = 16 \rightarrow x = ?$

Here, the operation is multiplication (Variable x is multiplied by 4.) and its inverse operation is division. To solve this equation, divide both sides of an equation by 4: $4x = 16 \rightarrow \frac{4x}{4} = \frac{16}{4} \rightarrow x = 4$.

Here is another example:

Solve this equation: $x + 8 = 0 \rightarrow x = ?$

In this equation, 8 is added to the variable x. The inverse operation of addition is subtraction. To solve this equation, subtract 8 from both sides of the equation: $x + 8 - 8 = 0 - 8$. Then: $x = -8$.

Let's do another example:

Solve this equation for x: $x - 12 = 0$.

Here, the operation is subtraction and its inverse operation is addition. To solve this equation, add 12 to both sides of the equation: $x - 12 + 12 = 0 + 12 \rightarrow x = 12$.

✏ Solve each equation.

23) $x - 3 = 6$

24) $5 - x = 7$

25) $2 = x + 2$

26) $-5 = 2 + b$

27) $-3 - x = 2$

28) $2x = 12$

29) $x + 5 = 0$

30) $32 = -8x$

31) $0 = 1 - a$

32) $\frac{3}{2}b = 9$

Multi−Step Equations

To solve a multi-step equation, we need to use a combination of inverse operations and the order of operations. The steps involve:

1. Combining like terms on one side of the equation. This step is important to simplify the equation and help us see the relationship between the variables and constants more clearly.

2. Bringing variables to one side by adding or subtracting. This step is important to isolate the variable on one side of the equation.

3. Simplifying the equation by using the inverse of addition or subtraction. This step is important to eliminate any constant terms and obtain the value of the variable.

4. Simplify further by using the inverse of multiplication or division. This step is important to obtain the value of the variable in the simplest form.

5. Checking the solution by plugging the value of the variable into the original equation. This step is important to confirm that the value found for the variable is a valid solution to the equation.

It is important to note that solving multi-step equations requires logical thinking, patience, and attention to detail, as well as a good understanding of mathematical operations and their inverse operations.

Let's give some examples to better understand the issue:

Solve this equation for x: $2x + 3 = 7x - 4$.

First, we'll bring all the x's to one side of the equation.

$$2x - 7x = -4 - 3 \rightarrow -5x = -7 \rightarrow x = \frac{7}{5}$$

Solve this equation for x: $3x - 2(x + 4) = 2x + 6$.

Expressions and Equations

First, we'll use the distributive property to expand $2(x + 4)$:

$3x - 2x - 8 = 2x + 6 \rightarrow x - 8 = 2x + 6$

Next, we'll bring all the x's to one side of the equation.

$-x = 14 \rightarrow x = -14$

Solve this equation for x: $4(x + 2) - 3x = 2x - 8$.

First, we'll use the distributive property to expand $4(x + 2)$:

$4x + 8 - 3x = 2x - 8 \rightarrow x + 8 = 2x - 8$

Next, we'll bring all the x's to one side of the equation.

$-x = -16 \rightarrow x = 16$

Solve this equation for x: $-5x + 4 = 24$.

Subtract 4 from both sides of the equation.

$-5x + 4 = 24 \rightarrow -5x + 4 - 4 = 24 - 4 \rightarrow -5x = 20$.

Divide both sides by -5, then:

$-5x = 20 \rightarrow \frac{-5x}{-5} = \frac{20}{-5} \rightarrow x = -4$.

Now, check the solution:

$x = -4 \rightarrow -5x + 4 = 24 \rightarrow -5(-4) + 4 = 24 \rightarrow 24 = 24$.

The answer $x = -4$ is correct.

✎ Solve each equation.

33) $1 = \frac{1}{2}x - 5$

34) $-(2 - x) = 3$

35) $-3 = -2x + 3$

36) $8 = 2(1 - 3x)$

37) $-\frac{1}{3}(2x - 10) = x$

38) $3(5x - 2) = 9$

39) $1 - (x + 1) = 2x + 3$

40) $3x - 1 = 5$

Rearrange Multi-Variable Equations

Rearranging a multi-variable equation involves isolating one variable on one side of the equation.

1. Determine the dependent variable: The variable that you want to solve for is called the dependent variable.
2. Isolate the dependent variable on both sides of the equation: To isolate the dependent variable, you need to use the inverse operations to undo the operations that are performed on it. This will leave the dependent variable on one side of the equation, and the other variables on the other side.
3. By undoing the operations on both sides, write the dependent variable on one side and the other variables on the other side of the equation. This step is important to get the equation in a form that we can solve for the dependent variable.

For example, if we have the equation $3x + 2y = 8$, we want to rearrange it to solve for x, we will isolate it on one side:

$$3x + 2y = 8 \rightarrow 3x = 8 - 2y \rightarrow x = \frac{(8 - 2y)}{3}$$

Another example, if we have the equation $x + 2y - 3z = 5$, we want to rearrange it to solve for y, we will isolate it on one side:

$$x + 2y - 3z = 5 \rightarrow 2y = -x + 3z + 5 \rightarrow y = \frac{(-x + 3z + 5)}{2}$$

It is important to note that rearranging an equation is a key step to solving a multi-variable equation, as it allows us to isolate one variable and find its value. Let's review another example:

Solve this equation for x: $\frac{1}{2}x - t = 5$.

To solve the equation "$\frac{1}{2}x - t = 5$" for x, we need to isolate x on one side of the equation. To do this, we add t to both sides of the equation: $\frac{1}{2}x = 5 + t$

Then we multiply both sides of the equation by 2: $x = 2(5 + t) = 10 + 2t$

Therefore, $x = 10 + 2t$ is the solution of the equation $\frac{1}{2}x - t = 5$.

Another example:

Solve for a in terms of b and c: $-c + b - a = 6$.

To solve the problem, find a. By undoing the operations on both sides, isolate a on one side of the equation. Add a to both sides of the equation. So,

$-c + b - a + a = 6 + a \rightarrow -c + b = a + 6$.

Now, subtract 6 from both sides of the equation, then:

$-c + b - 6 = a + 6 - 6 \rightarrow -c + b - 6 = a$.

In this case, the above equation in terms of a becomes: $a = -c + b - 6$.

Let's do another example:

Solve $V = \frac{1}{3}\pi r^2 h$ for h.

To isolate h on both sides of the equation, just multiply the sides by the expression $\frac{3}{\pi r^2}$. Therefore,

$\frac{3}{\pi r^2} \times V = \frac{3}{\pi r^2}\left(\frac{1}{3}\pi r^2 h\right) \rightarrow h = \frac{3V}{\pi r^2}$

✏ Rearrange Multi-Variable Equations.

41) $\frac{1}{2}qp - l = w$ for q.

42) $w + z = \frac{5x}{2y}$ for y.

43) $T = \frac{PV}{nR}$ for P.

44) $d + a = -b - c$ for c.

Finding Midpoint

The midpoint of a line segment is a point that is equidistant from the two endpoints of the line segment. It is the point that is exactly in the middle of the line segment.

The midpoint of a line segment can be found by averaging the x-coordinates and the y-coordinates of the endpoints of the line segment.

The midpoint of two endpoints $A(x_1, y_1)$ and $B(x_2, y_2)$ can be found using this formula: $M\left(\frac{x_1+x_2}{2}, \frac{y_1+y_2}{2}\right)$.

Let's review some examples:

Find the midpoint of the line segment with the given endpoints: $(5, -2), (3, 6)$.

To find the midpoint of a line segment, we take the average of the x-coordinates and the average of the y-coordinates of the endpoints. Given endpoints $(5, -2)$ and $(3, 6)$. The midpoint is: $\left(\frac{5+3}{2}, \frac{-2+6}{2}\right) = (4, 2)$

So, the midpoint of the line segment with endpoints $(5, -2)$ and $(3, 6)$ is $(4, 2)$.

Another example:

Find the midpoint of the line segment with the given endpoints: $(3, 4)$ and $(7, -2)$.

The midpoint is: $\left(\frac{3+7}{2}, \frac{4+(-2)}{2}\right) = (5, 1)$.

Let's review another example:

One endpoint of a line segment is $(3, 4)$ and its midpoint is $(6, -4)$. Find the other endpoint.

We know that the midpoint of a line segment is the average of the x and y coordinates of its endpoints.

Expressions and Equations

Given that one endpoint of a line segment is (3,4) and its midpoint is (6,−4), we can use this information to find the other endpoint.

If (x_1, y_1) is one endpoint and (x_2, y_2) is the other endpoint, then the midpoint is $\left(\frac{x_1+x_2}{2}, \frac{y_1+y_2}{2}\right)$.

So, in this case, the coordinates of the other endpoint are:

$x_2 = (2 \times 6) - 3 = 9$, $y_2 = (2 \times (-4)) - 4 = -12$.

So, the other endpoint of the line segment is (9,−12).

Let's solve another example together:

Find the midpoint of the line segment with the given endpoints: (2,−4), (6,8).

Midpoint $= \left(\frac{x_1+x_2}{2}, \frac{y_1+y_2}{2}\right) \rightarrow (x_1, y_1) = (2,-4)$ and $(x_2, y_2) = (6,8)$.

Midpoint $= \left(\frac{2+6}{2}, \frac{-4+8}{2}\right) \rightarrow \left(\frac{8}{2}, \frac{4}{2}\right) \rightarrow M(4,2)$.

Another example:

Find the midpoint of the line segment with the given endpoints: (−2,3), (6,−7).

Midpoint $= \left(\frac{x_1+x_2}{2}, \frac{y_1+y_2}{2}\right) \rightarrow (x_1, y_1) = (-2,3)$ and $(x_2, y_2) = (6,-7)$.

Midpoint $= \left(\frac{-2+6}{2}, \frac{3+(-7)}{2}\right) \rightarrow \left(\frac{4}{2}, \frac{-4}{2}\right) \rightarrow M(2,-2)$.

✏ Find the midpoint of the line segment with the given endpoints.

45) (3,2) and (3,6)

46) (3,5) and (0,1)

47) (0,2) and (−1,3)

48) (4,8) and (8,3)

49) (1,3) and (−3,−1)

50) (−1,−3) and (2,−5)

51) (1,4) and (−1,−1)

52) (1,1) and (5,1)

Finding the Distance of Two Points

The distance between two points is the length of the shortest path between them, and it can be represented by a straight-line segment connecting the two points. The most common method for finding the distance between two points is the distance formula. The distance formula is used to find the distance between two points in a coordinate plane. The formula is based on the Pythagorean theorem, which states that in a right triangle, the square of the length of the hypotenuse is equal to the sum of the squares of the lengths of the other two sides.

Use the following formula to find the distance of two points with the coordinates $A(x_1, y_1)$ and $B(x_2, y_2)$:

$$d = \sqrt{(x_2 - x_1)^2 + (y_2 - y_1)^2}$$

Where d is the distance between the two points, (x_1, y_1) and (x_2, y_2) are the coordinates of the two points.

For example, if the coordinates of two points are $(3,4)$ and $(6,8)$, the distance between the two points can be found by applying the distance formula:

$$d = \sqrt{(6-3)^2 + (8-4)^2} = \sqrt{(3^2 + 4^2)} = \sqrt{(9+16)} = \sqrt{25} = 5$$

It is important to note that the distance formula works for any two points in a plane, whether they are in a two-dimensional Cartesian coordinate system or a three-dimensional space. The distance formula can also be used for other forms of measurements, such as distance in physics, in which case the formula may have different variables.

Another example:

Find the distance between $(4,2)$ and $(-5,-10)$ on the coordinate plane.

Expressions and Equations

Use the distance of two points formula: $d = \sqrt{(x_2 - x_1)^2 + (y_2 - y_1)^2}$.

Considering that: $(x_1, y_1) = (4,2)$ and $(x_2, y_2) = (-5, -10)$.

Then: $d = \sqrt{(-5-4)^2 + (-10-2)^2} = \sqrt{(-9)^2 + (-12)^2} = \sqrt{81 + 144} = \sqrt{225} = 15$

Then: $d = 15$.

Consider another example:

Find the distance of two points $(-1,5)$ and $(-4,1)$.

Use the distance of two points formula: $d = \sqrt{(x_2 - x_1)^2 + (y_2 - y_1)^2}$.

Since $(x_1, y_1) = (-1,5)$, and $(x_2, y_2) = (-4,1)$.

Then: $d = \sqrt{(-4 - (-1))^2 + (1-5)^2} = \sqrt{(-3)^2 + (-4)^2} = \sqrt{9 + 16} = \sqrt{25} = 5$.

Then: $d = 5$.

Let's solve another example together:

Find the distance between $(-6,5)$ and $(-1,-7)$.

Use the distance of two points formula: $d = \sqrt{(x_2 - x_1)^2 + (y_2 - y_1)^2}$.

According to: $(x_1, y_1) = (-6,5)$ and $(x_2, y_2) = (-1,-7)$. Then:

$$d = \sqrt{(-1-(-6))^2 + (-7-5)^2} = \sqrt{(5)^2 + (-12)^2} = \sqrt{25 + 144} = \sqrt{169} = 13$$

✎ Find the distance between each pair of points.

53) $(-6,9)$, and $(-6,8)$

54) $(-1,3)$, and $(3,3)$

55) $(-2,0)$, and $(0,2)$

56) $(-5,0)$, and $(-2,-4)$

57) $(2,-5)$, and $(6,-4)$

58) $(3,10)$, and $(8,3)$

59) $(3,1)$, and $(9,1)$

60) $(-1,3)$, and $(1,0)$

61) $(-1,3)$, and $(3,3)$

62) $(0,-5)$, and $(0,-1)$

Graphing Absolute Value Equations

The general form of an absolute value function (that is linear) is:

$$y = |mx + b| + c$$

The vertex (the lowest or the highest point) is located at $\left(\frac{-b}{m}, c\right)$.

A vertical line that divides the graph into two equal halves is: $x = -\frac{b}{m}$.

To graph an absolute value equation, you can find the vertex and some other points by substituting some values for x and solving for y. Once you have the vertex and a few other points, you can use them to sketch the graph of the function.

For example, if you have the equation: $y = |2x + 1| + 1$, the vertex is located at $\left(\frac{-1}{2}, 1\right)$.

To graph the equation, you can substitute $x = -1$, $x = 0$, and $x = 1$ into the equation and solve for y to get 2, 2, and 4, respectively.

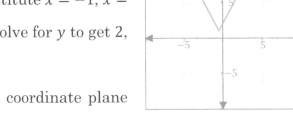

These points can be plotted on the coordinate plane and connected to sketch the graph of the function.

It is important to note that the graph of an absolute value function is a *V*-shaped graph and it is different from a linear function graph which is a straight line.

Let's do an example:

Graph $y = |x + 2|$.

Find the vertex $\left(\frac{-b}{m}, c\right)$.

According to the general form of an absolute value function: $y = |mx + b| + c$.

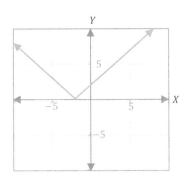

Expressions and Equations

We have: $x = \frac{-b}{m} \rightarrow x = \frac{-2}{1} = -2$.

The point $(-2, 0)$ is the vertex of the graph and represents the center of the table of values. Create the table and plot the ordered pairs.

Now, find the points and graph the equation.

x	$y = \|x + 2\|$
-4	$\|-4 + 2\| = 2$
-3	$\|-3 + 2\| = 1$
-2	$\|-2 + 2\| = 0$
-1	$\|-1 + 2\| = 1$

Let's do an example:

Graph $y = |x - 3|$.

Find the vertex $\left(\frac{-b}{m}, c\right)$.

According to the general form of an absolute value function: $y = |mx + b| + c$.

We have: $x = \frac{-b}{m} \rightarrow x = \frac{-(-3)}{1} = 3$.

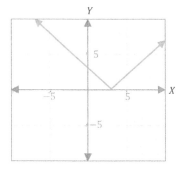

The point $(3, 0)$ is the vertex of the graph and represents the center of the table of values. Create the table and plot the ordered pairs.

Now, find the points and graph the equation.

x	$y = \|x - 3\|$
1	$\|1 - 3\| = 2$
2	$\|2 - 3\| = 1$
3	$\|3 - 3\| = 0$
4	$\|4 - 3\| = 1$

✏️ Graphing absolute value equations.

63) $y = -|x| - 1$

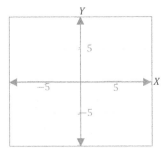

64) $y = -|x - 3|$

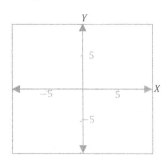

Chapter 3: Answers

1) $4x + 2$
2) $-2x + 2$
3) $7x - 4$
4) $8x - 9$
5) $-9x + 5$
6) $-3x - 4$
7) $2x - 12$
8) $-x^2 - 3x + 6$
9) 5
10) 4
11) 3
12) 1
13) 0
14) 17
15) -2
16) -3
17) 1
18) -1
19) 0
20) -2
21) -10
22) -1
23) 9
24) -2
25) 0
26) -7
27) -5
28) 6
29) -5
30) -4
31) 1
32) 6
33) 12
34) 5
35) 3
36) -1
37) 2
38) 1
39) -1
40) 2
41) $q = \frac{2(w+l)}{p}$
42) $y = \frac{5x}{2(w+z)}$
43) $P = \frac{nTR}{V}$
44) $c = -b - a - d$
45) $(3,4)$
46) $\left(\frac{3}{2}, 3\right)$
47) $\left(-\frac{1}{2}, \frac{5}{2}\right)$
48) $\left(6, \frac{11}{2}\right)$
49) $(-1,1)$
50) $\left(\frac{1}{2}, -4\right)$
51) $\left(0, \frac{3}{2}\right)$

52) (3,1)

53) 1

54) 4

55) $2\sqrt{2}$

56) 5

57) $\sqrt{17}$

58) $\sqrt{74}$

59) 6

60) $\sqrt{13}$

61) 4

62) 4

63) $y = -|x| - 1$

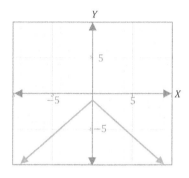

64) $y = -|x - 3|$

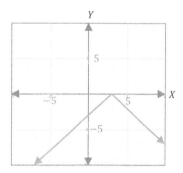

CHAPTER
4 Linear Functions

Math topics that you'll learn in this chapter:

- ☑ Finding Slope
- ☑ Writing Linear Equations
- ☑ Graphing Linear Inequalities
- ☑ Finding the Slope, x −intercept, and y −intercept
- ☑ Write an Equation from a Graph
- ☑ Slope-intercept Form and Point-slope Form
- ☑ Write a Point-slope Form Equation from a Graph
- ☑ Find x −and y −intercepts in the Standard Form of Equation
- ☑ Graph an Equation in the Standard Form
- ☑ Equations of Horizontal and Vertical Lines
- ☑ Graph a Horizontal or Vertical Line
- ☑ Graph an Equation in Point-Slope Form
- ☑ Equation of Parallel or Perpendicular Lines
- ☑ Compare Linear Function's Graph and Equations
- ☑ Two-variable Linear Equations Word Problems

Finding Slope

The slope of a line is a measure of its steepness and direction on a coordinate plane. A coordinate plane is a two-dimensional surface made up of two perpendicular number lines, the horizontal x −axis and the vertical y −axis. The point where the two axes intersect is called the origin. An ordered pair (x, y) represents the coordinates of a point on the plane.

A line on a coordinate plane is represented by connecting two points. To determine the slope of a line, we need either the equation of the line or two points on the line.

The slope of a line with two points $A(x_1, y_1)$ and $B(x_2, y_2)$ can be found by using this formula: $\frac{y_2 - y_1}{x_2 - x_1} = \frac{rise}{run}$, which is the ratio of the rise (the change in y) to the run (the change in x).

The equation of a line is typically written as $y = mx + b$ where m is the slope and b is the y −intercept.

Let's review an example:

Find the slope of the line through these two points: $A(3, -6)$ and $B(-2, 2)$.

The slope of a line is the ratio of the change in the y −coordinate to the change in the x −coordinate between two points on the line. To find the slope of a line through two points, we use the formula: Slope $= \frac{y_2 - y_1}{x_2 - x_1}$

Given two points $A(3, -6)$ and $B(-2, 2)$. Slope $= \frac{2 - (-6)}{-2 - 3} = \frac{8}{-5} = -1.6$

So, the slope of the line through these two points $A(3, -6)$ and $B(-2, 2)$ is -1.6.

Another example: Find the slope of the line with equation $y = -3x + 8$.

Linear Functions

The slope of a line is the coefficient of the x-term in the equation of the line. In the equation $y = -3x + 8$, the coefficient of the x-term is -3. So, the slope of the line is -3.

One more example: Find the slope of the line through these two points: $A(4,0)$ and $B(8,12)$.

The slope of a line is the ratio of the change in the y-coordinate to the change in the x-coordinate between two points on the line. To find the slope of a line through two points, we use the formula: Slope $= \frac{y_2 - y_1}{x_2 - x_1}$

Given two points $A(4,0)$ and $B(8,12)$. Slope $= \frac{12-0}{8-4} = \frac{12}{4} = 3$

So, the slope of the line through these two points $A(4,0)$ and $B(8,12)$ is 3.

To better explain the problem, I can draw your attention to this example:

Find the slope of the line through these two points: $A(1,-6)$ and $B(3,2)$.

Use the formula: Slope $= \frac{y_2 - y_1}{x_2 - x_1}$. Let (x_1, y_1) be $A(1,-6)$ and (x_2, y_2) be $B(3,2)$.

(Remember, you can choose any point for (x_1, y_1) and (x_2, y_2)).

Then: Slope $= \frac{y_2 - y_1}{x_2 - x_1} = \frac{2-(-6)}{3-1} = \frac{8}{2} = 4$.

The slope of the line through these two points is 4.

✎ Find the slope of each line.

1) $y = x - 5$

2) $y = 2x + 6$

3) $y = -5x - 8$

4) Line through $(2,6)$ and $(5,0)$

5) Line through $(8,0)$ and $(-4,3)$

6) Line through $(-2, -4)$ and $(-4, 8)$

Writing Linear Equations

The equation of a line in slope-intercept form is $y = mx + b$. This form of the equation is particularly useful because it allows us to quickly identify the slope (m) and the $y-$intercept (b) of the line.

To write the equation of a line, the first step is to identify the slope. The slope is the ratio of the change in y to the change in x between any two points on the line. It can also be represented by the letter "m" in the equation $y = mx + b$.

The next step is to find the $y-$intercept. The $y-$intercept is the point at which the line crosses the $y-$axis. It can be represented by the letter "b" in the equation $y = mx + b$. To find the $y-$intercept, we can substitute the slope and the coordinates of a point (x, y) on the line into the equation.

For example, if we know a line has a slope of 3 and the point $(2,5)$ is on the line, we can substitute these values into the equation $y = mx + b$: $y = 3x + b$, and $5 = 3(2) + b \to 5 = 6 + b \to b = -1$. So, the equation of the line is $y = 3x - 1$.

It is important to note that the slope-intercept form of the equation of a line is just one of the ways to represent a line, and it can be transformed into other forms such as point-slope form or standard form.

Example: What is the equation of the line that passes through $(2, -4)$ and has a slope of 3?

The equation of a line can be written in the form $y = mx + b$, where m is the slope of the line and b is the $y-$intercept (The point at which the line crosses the $y-$axis). To find the equation of a line that passes through a given point and has a given slope, we can use the point-slope form of a line.

The point-slope form of a line is: $y - y_1 = m(x - x_1)$. where (x_1, y_1) is a point on the line and m is the slope of the line. Given point $(2, -4)$ and slope 3, $y - (-4) = 3(x - 2) \to y = 3x - 6 - 4 \to y = 3x - 10$. So, the equation of the line that passes through $(2, -4)$ and has a slope of 3 is $y = 3x - 10$.

Another example: What is the equation of the line that passes through (5,2) and has a slope of 4?

The general slope-intercept form of the equation of a line is: $y = mx + b$, where m is the slope and b is the y-intercept.

By substitution of the given point and given slope: $y = mx + b \rightarrow 2 = (4)(5) + b$.

So, $b = 2 - 20 = -18$, and the required equation of the line is: $y = 4x - 18$.

One more example: Write the equation of the line through two points $A(6,8)$ and $B(-1,22)$.

First, find the slope: slope $= \frac{y_2 - y_1}{x_2 - x_1} = \frac{22-8}{-1-6} = \frac{14}{-7} = -2 \rightarrow m = -2$.

To find the value of b, use either point and plug in the values of x and y in the equation. The answer will be the same: $y = -2x + b$. Let's check both points.

Then: $(6,8) \rightarrow y = mx + b \rightarrow 8 = -2(6) + b \rightarrow b = 20$.

$(-1,22) \rightarrow y = mx + b \rightarrow 22 = -2(-1) + b \rightarrow b = 20$.

The y-intercept of the line is 4. The equation of the line is: $y = -2x + 20$.

✏ Solve.

7) What is the equation of a line with slope 4 and intercept 16?

8) What is the equation of a line with slope 3 and passes through point (1,5)?

9) What is the equation of a line with slope -5 and passes through point $(-2,7)$?

10) The slope of a line is -4 and it passes through point $(-6,2)$. What is the equation of the line?

11) The slope of a line is -3 and it passes through point $(-3,-6)$. What is the equation of the line?

Graphing Linear Inequalities

To graph a linear inequality, we first need to graph the "equals" line. This is the line that separates the solution of the inequality from the non-solution. The equation of this line can be written in slope-intercept form, point-slope form, or standard form.

Once we have the equation of the line, we can graph it on the coordinate plane. The type of inequality symbol used determines the type of line we use to graph the equation. For less than (<) and greater than (>) signs, we use a dashed line. For less than or equal to (≤) and greater than or equal to (≥) we use a solid line. The next step is to choose a testing point. This can be any point on both sides of the line. We then substitute the values of x and y of that point into the inequality. If the inequality is true, then that part of the line is the solution. If the inequality is not true, then the other part of the line is the solution.

It is important to note that when we graph a linear inequality, we graph the equation of the line, and then shade the half-plane that contains the solution of the inequality, depending on the inequality symbol.

Let's review an example: Sketch the graph of inequality: $y < 2x + 4$.

To sketch the graph of the inequality $y < 2x + 4$, we first need to graph the "equals" line which is $y = 2x + 4$.

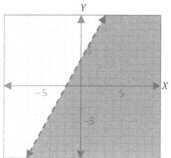

1. First, we need to find two points that are on the line $y = 2x + 4$. We can do this by substituting $x = -1$ and $x = 1$ and solving for y.

$$x = -1 \rightarrow y = 2(-1) + 4 = 2$$
$$x = 1 \rightarrow y = 2(1) + 4 = 6$$

So, two points that are on the line are $(-1, 2)$ and $(1, 6)$.

Linear Functions

2. Next, we plot these points on the coordinate plane and draw the line that passes through them.
3. We use a dashed line because the inequality is "less than" and the solution is the area below the line.
4. Shade the region below the line to represent the solution set of the inequality.
5. To check if our graph is correct, we can pick a point in the shaded area, say $(-2, -2)$, and substitute it into the inequality. It should be true, as $-2 < 2 \times (-2) + 4$.

Another example: Sketch the graph of inequality: $y > 3x + 1$.

To draw the graph of $y > 3x + 1$, you first need to graph the line: $y = 3x + 1$.

Since there is a less than ($>$) sign, draw a dashed line. The slope is 3 and the y-intercept is 1.

Then, choose a testing point and substitute the value of x and y from that point into the inequality. The easiest point to test is the origin: $(0,0) \rightarrow 0 > 3(0) + 1 \rightarrow 0 > 1$.

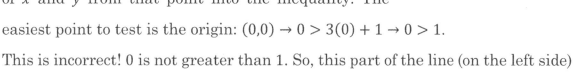

This is incorrect! 0 is not greater than 1. So, this part of the line (on the left side) is the solution of this inequality.

✎ Sketch the graph of each linear inequality.

12) $y > 4x + 2$

13) $y < -2x + 5$

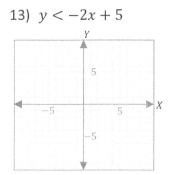

Finding the Slop, x–intercept and y–intercept

When a linear function is graphed, it will always be a straight line. The slope-intercept form of a linear equation is $y = mx + b$, where m is the slope of the line and b is the y–intercept. The slope of a line is a measure of how steep the line is and it is defined as the ratio of the change in the y–coordinate to the change in the x–coordinate between two points on the line. The y–intercept is the point at which the line crosses the y–axis, which means when $x = 0$ the y–coordinate is equal to b. In this form, m and b are constants and x and y are variables. By understanding the slope and y–intercept, we can fully describe the linear function and its characteristics.

In addition to the y–intercept and slope, the x–intercept is also an important aspect of a linear function. The x–intercept is the point at which the line crosses the x–axis, which means when $y = 0$, the x–coordinate is equal to $\frac{-b}{m}$. By finding both x–intercept and y–intercept we can plot the line on the cartesian plane.

Knowing the equation of a line in slope-intercept form allows you to easily graph the line and understand its properties, such as its direction, whether it is increasing or decreasing, and its overall shape.

Let's review an example: Find the slope, x and y intercepts for the following equation. $2x + 8y = 12$

The slope-intercept form of a linear equation is $y = mx + b$, where m is the slope of the line and b is the y–intercept. However, the equation given, $2x + 8y = 12$, is not in that form. To find the slope, x–intercept, and y–intercept, we need to manipulate the equation into that form.

To find the slope, we need to get the equation in the form $y = mx + b$, so we can read off the value of m. To do this, we can divide both sides of the

Linear Functions

equation by 8: $\frac{2x}{8} + \frac{8y}{8} = \frac{12}{8} \rightarrow \frac{x}{4} + y = \frac{3}{2} \rightarrow y = -\frac{x}{4} + \frac{3}{2}$. So, the slope of the line is $m = -\frac{1}{4}$. To find the y-intercept we need to put $x = 0$ in the equation.

$$2 \times 0 + 8y = 12 \rightarrow 8y = 12 \rightarrow y = \frac{12}{8} \rightarrow y = \frac{3}{2}$$

So, the y-intercept of the line is $(0, \frac{3}{2})$. To find the x-intercept we need to put $y = 0$ in the equation: $2x + 8 \times 0 = 12 \rightarrow 2x = 12 \rightarrow x = 6$.

So, the x-intercept of the line is $(6, 0)$. To sum up, the slope of the line is $-\frac{1}{4}$, the y-intercept is $(0, \frac{3}{2})$ and the x-intercept is $(6, 0)$.

Another example: Find the slope, x and y intercepts for the following equation.

$$x + 3y = 15$$

To find the x-intercept, arrange the equation $x + 3y = 15$ so that x is isolated:

$$x = -3y + 15$$

Using the point-slope formula, we see that the x-intercept is 15.

To find the y-intercept, rearrange the equation $x + 3y = 15$ so that y is isolated:

$$x + 3y = 15 \rightarrow 3y = -x + 15 \rightarrow y = \frac{-1}{3}x + 5.$$

So, the y-intercept is 5.

To find the slope, use this formula: $y = mx + b \rightarrow y = \frac{-1}{3}x + 5$. The slope is $-\frac{1}{3}$.

✍ Find the slope, x and y intercepts for the following equations.

14) $y = -\frac{1}{3}x - 1$

   y-intercept: _____

   x-intercept: _____

   Slope: _____

15) $y = 5x + 10$

   y-intercept: _____

   x-intercept: _____

   Slope: _____

Write an Equation from a Graph

To write an equation of a line in slope-intercept form, given a graph of that equation, we need to pick two points on the line and use them to find the slope. The slope is the ratio of the change in y to the change in x between any two points on the line. It is also the value of "m" in the equation $y = mx + b$.

Once we have the slope, we can find the coordinates of the y-intercept. The y-intercept is the point at which the line crosses the y-axis. It should be of the form $(0, b)$ where b is the y-coordinate. The b is the value of "b" in the equation $y = mx + b$.

We can now write the equation by substituting the numerical values for m and b we found. The final equation will be of the form $y = mx + b$ where m is the slope and b is the y-intercept.

To check if the equation is correct, we can pick a point on the line (not the y-intercept) and substitute its coordinates (x, y) into the equation. If the equation is true for that point, then the equation is correct.

Let's review an example:

Write the equation of the following line in slope-intercept form.

First, pick two points on the line for example, $(-3, 3)$ and $(6, 0)$.

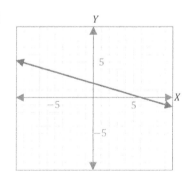

Use these points to calculate the slope:

$$m = \frac{0-3}{6-(-3)} = \frac{-3}{9} = -\frac{1}{3}.$$

Next, find the y-intercept: $(0, 2)$. Thus, $b = 2$.

Therefore, the equation of this line is:

$$y = -\frac{1}{3}x + 2.$$

Now, let's check the equation by picking another point on the line. Let's choose point (3,1).

Then: $(3,1) \to y = -\frac{1}{3}x + 2 \to 1 = -\frac{1}{3}(3) + 2 \to 1 = -1 + 2$. This is true!

Another example: Write the equation of the following line in slope-intercept form.

First, pick two points on the line for example, (2,1) and (3,3).

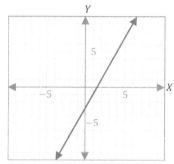

Use these points to calculate the slope: $m = \frac{3-1}{3-2} = \frac{2}{1} = 2$.

Next, find the y–intercept: $(0, -3)$. Thus, $b = -3$.

Therefore, the equation of this line is: $y = 2x - 3$.

Now, let's check the equation by picking another point on the line. Let's choose point $(-2, -7)$.

Then: $(-2, -7) \to y = 2x - 3 \to -7 = 2(-2) - 3 \to -7 = -4 - 3$. This is true!

✎ Write an equation of each of the following lines in slope-intercept from.

16) _____

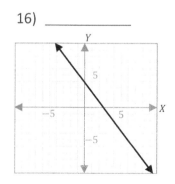

17) _____

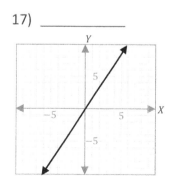

Slope-intercept Form and Point-slope Form

The slope-intercept form is a way to express the equation of a line. The equation of a line in slope-intercept form is written as $y = mx + b$, where m is the slope of the line and b is the $y-$intercept (the point where the line crosses the $y-$axis). This form is useful because it allows you to easily determine the slope and $y-$intercept of a line from its equation.

The point-slope form is another way to express the equation of a line. The equation of a line in point-slope form is written as $y - y_1 = m(x - x_1)$, where (x_1, y_1) is a point on the line and m is the slope of the line. This form is useful because it allows you to easily find the equation of a line when you know the slope and a point on the line.

Let's review an example:

Find the equation of a line with point (3,7) and slope -4, and write it in slope-intercept and point-slope forms.

Slope-intercept form: $y = -4x + b$. By using the point (3,7) we can find:

$b = y - mx = 7 + 4 \times 3 = 7 + 12 = 19$.

So, the equation will be $y = -4x + 19$. Point-slope form: $y - 7 = -4(x - 3)$.

Another example:

Find the equation of a line with point $(2, -2)$ and slope 2, and write it in slope-intercept and point-slope forms.

Slope-intercept form: $y = 2x + b$. By using the point $(2, -2)$ we can find:

$b = y - mx = -2 - 2 \times 2 = -2 - 4 = -6$

So, the equation will be $y = 2x - 6$. Point-slope form: $y + 2 = 2(x - 2)$.

One more example:

Find the equation of a line with point (0,4) and slope 3, and write it in slope-intercept and point-slope forms.

Slope-intercept form: $y = 3x + b$. By using the point (0,4) we can find:

$$b = y - mx = 4 - 3 \times 0 = 4$$

So, the equation will be $y = 3x + 4$. Point-slope form: $y - 4 = 3(x - 0)$.

To better explain the problem, I can draw your attention to this example:

Find the equation of a line with point (4,6) and slope 3, and write it in slope-intercept and point-slope forms.

For point-slope form, we have the point and slope: $x_1 = 4$, $y_1 = 6$, $m = 3$. Then:

$$y - y_1 = m(x - x_1) \rightarrow y - 6 = 3(x - 4)$$

The slope-intercept form of a line is: $y = mx + b$.

Since $y = 6$, $x = 4$, $m = 3$, we just need to solve for b.

$$y = mx + b \rightarrow 6 = 3(4) + b \rightarrow b = -6$$

Slope-intercept form: $y = 3x - 6$.

Another example: Find the equation of a line with point (1,5) and slope -3, and write it in slope-intercept and point-slope forms.

For point-slope form, we have the point and slope: $x_1 = 1$, $y_1 = 5$, and $m = -3$.
Then: $y - y_1 = m(x - x_1) \rightarrow y - 5 = -3(x - 1)$.

The slope-intercept form of a line is: $y = mx + b$.

Since $y = 5$, $x = 1$, $m = -3$, we just need to solve for b.

$$y = mx + b \rightarrow 5 = -3(1) + b \rightarrow b = 8$$

Slope-intercept form: $y = -3x + 8$.

✎ Find the equation of each line.

18) Through: $(6, -6)$, slope $= -2$

   Point-slope form: _____

   Slope-intercept form: _____

19) Through: $(-7, 7)$, slope $= 4$

   Point-slope form: _____

   Slope-intercept form: _____

Write a Point-slope Form Equation from a Graph

The point-slope form equation is used to find the equation of a straight line when given the slope and a point on the line. This form of the equation is only useful when the slope and a point on the line are known. The equation is written as $y - y_1 = m(x - x_1)$, where (x_1, y_1) is a point on the line and m is the slope. The variable (x, y) in the equation represents a random point on the line.

To find the equation of a straight line using the point-slope form, you can follow these steps:

1. Determine the slope, m, of the line using the slope formula: $m = \frac{change\ in\ y}{change\ in\ x}$. Then, find the coordinates (x_1, y_1) of a point on the line.

2. Substitute the slope and point coordinates into the point-slope formula: $y - y_1 = m(x - x_1)$.

3. Simplify the equation to get the standard form of the line's equation.

Let's review an example: According to the following graph, what is the equation of the line in point-slope form?

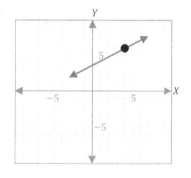

First, you should find the slope of the line (m). The coordinate of the red point is $(4,6)$. Consider another random point on the line such as $(7,8)$. Put this value in the slope formula: $m = \frac{change\ in\ y}{change\ in\ x} = \frac{8-6}{7-4} = \frac{2}{3} \rightarrow m = \frac{2}{3}$.

Now, write the equation in point-slope form using the coordinate of the red point is $(4,6)$ and $m = \frac{2}{3}$:

$$y - y_1 = m(x - x_1) \rightarrow y - 6 = \frac{2}{3}(x - 4).$$

Therefore, the equation of the line in point-slope form is $y - 6 = \frac{2}{3}(x - 4)$.

Linear Functions

Another example: Find the equation of the graph below in point-slope form. Consider two arbitrary points like $(-5,2)$ and $(2,6)$ on the line graph. (marked points on the line graph). Then, we calculate the slope of the line passing through these two points such as (x_1, y_1) and (x_2, y_2) using the slope formula: $m = \frac{y_2 - y_1}{x_2 - x_1}$. For the specified points, we have: $m = \frac{6-2}{2-(-5)} = \frac{4}{7}$. In the next step,

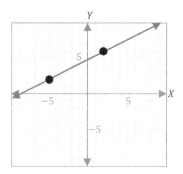

substitute an arbitrary point on the line like $(2,6)$ and the slope obtained $\frac{4}{7}$ in the point-Slope formula $y - y_1 = m(x - x_1)$. Therefore, the equation of the line in point-slope form is $y - 6 = \frac{4}{7}(x - 2)$.

Find the equation of each line.

20) _____

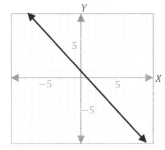

21) _____

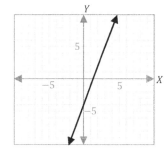

22) _____

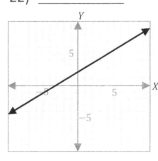

23) _____

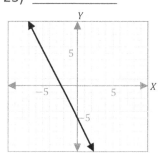

Find $x-$ and $y-$intercepts in the Standard Form of Equation

The standard form of a linear equation, also known as the general form, is written as $Ax + By = C$. In this equation, A, B, and C are integers, and x and y are the variables.

To convert a linear equation to standard form, you can rearrange the equation so that it is in the form $Ax + By = C$. It's important to note that the coefficients A, B, and C should be integers and the variables should be in the proper order as mentioned in the standard form equation.

The $x-$intercept of a line is the point where the line intersects the $x-$axis. At this point, the value of y is equal to zero. The $y-$intercept of a line is the point where the line intersects the $y-$axis. At this point, the value of x is equal to zero.

Let's review an example: Find the $x-$ and $y-$intercepts of the line $6x + 4y = 12$.

To find the $x-$intercept, set y equal to 0 and solve for x:

$6x + 4(0) = 12 \rightarrow 6x = 12 \rightarrow x = 2$

The $x-$intercept is $(2,0)$.

To find the $y-$intercept, set x equal to 0 and solve for y:

$6(0) + 4y = 12 \rightarrow 4y = 12 \rightarrow y = 3$

The $y-$intercept is $(0,3)$.

The $x-$intercept is $(2,0)$ and the $y-$intercept is $(0,3)$.

Another example:

Find the $x-$ and $y-$intercepts of the line $3x - 6y = 0$.

To find the $x-$intercept, set y equal to 0 and solve for x:

$3x - 6(0) = 0 \rightarrow 3x = 0 \rightarrow x = 0$

The x −intercept is $(0,0)$.

To find the y −intercept, set x equal to 0 and solve for y:

$3(0) - 6y = 0 \to -6y = 0 \to y = 0$

The y −intercept is $(0,0)$.

Finally, the x − and y −intercept is the origin of the coordinate.

One more example: Find the x − and y −intercepts of the line $2x + 4y = 8$.

To find the x −intercept, set y equal to 0 and solve for x:

$2x + 4(0) = 8 \to 2x = 8 \to x = 4$

The x −intercept is $(4,0)$.

To find the y −intercept, set x equal to 0 and solve for y:

$2(0) + 4y = 8 \to 4y = 8 \to y = 2$

The y −intercept is $(0,2)$.

The x − and y −intercepts are $(4,0)$ and $(0,2)$, respectively.

For a better understanding, let's solve another example together:

Find the x − and y −intercepts of line $-8x + 16y = 64$.

To find the x −intercept, you can consider y equal to 0 and solve for x:

$-8x + 16y = 64 \to -8x + 16(0) = 64 \to -8x = 64 \to x = -8$

The x −intercept is -8.

To find the y −intercept, you can consider x equal to 0 and solve for y:

$-8x + 16y = 64 \to -8(0) + 16y = 64 \to 16y = 64 \to y = 4$.

The y −intercept is 4.

✎ Find the x −intercept of each line.

24) $21x - 3y = -18$

25) $20x + 20y = -10$

26) $8x + 6y = 16$

27) $2x - 4y = -12$

Graph an Equation in the Standard Form

To graph an equation in standard form, you can follow these steps:

1. Plot the $x-$intercept and $y-$intercept. You can find the $x-$intercept by setting y equal to 0 and solving for x, and you can find the $y-$intercept by setting x equal to 0 and solving for y. Using the slope-intercept form, find the slope of the line by solving for m from the standard form equation.

2. Using the slope found in step 1, pick a point on the line other than the $x-$intercept or $y-$intercept, and use the slope to find another point on the line.

3. Connect the dots you've plotted to form a straight line.

For example, to graph the equation $2x - 4y = 8$:

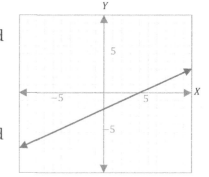

1. To find the $x-$intercept, set y equal to 0 and solve for x: $2x - 4(0) = 8 \rightarrow x = 4$;

   The $x-$intercept is $(4,0)$.

2. To find the $y-$intercept, set x equal to 0 and solve for y: $2(0) - 4y = 8 \rightarrow y = -2$;

   The $y-$intercept is $(0,-2)$.

3. We can find the slope by dividing both sides by -4, then we have $y = \frac{1}{2}x - 2$, the slope is $\frac{1}{2}$.

4. Using the slope found in step 3, we can pick a point on the line other than the $x-$intercept or $y-$intercept, for example $(6,1)$, we can use the slope to find another point on the line $(-4,-4)$, and connect them by a straight line to form a graph.

Another example:

Graph the equation $4x + 2y = 12$:

1. To find the $x-$intercept, set y equal to 0 and solve for x: $4x + 2(0) = 12 \to x = 3$;
   The $x-$intercept is $(3,0)$.

2. To find the $y-$intercept, set x equal to 0 and solve for y: $4(0) + 2y = 12 \to y = 6$;
   The $y-$intercept is $(0,6)$.

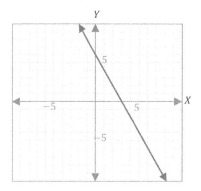

3. We can find the slope by dividing both sides by 4, then we have $y = -2x + 6$, and the slope is -2.

4. Using the slope found in step -2, we can pick a point on the line other than the $x-$intercept or $y-$intercept, for example $(2,2)$, we can use the slope to find another point on the line $(4,-2)$, and connect them by a straight line to form a graph.

In this example, we can see that the $x-$intercept is $(3,0)$, the $y-$intercept is $(0,6)$, and the slope of the line is -2. By plotting the $x-$intercept, $y-$intercept, and any point on the line, we can graph the equation on a coordinate plane.

✎ Graph each equation.

28) $4x - 5y = 40$

29) $9x - 8y = -72$

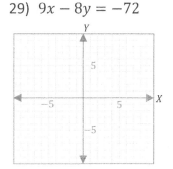

Equations of Horizontal and Vertical Lines

The equation of a horizontal line is of the form $y = b$, where b is a constant. This is because a horizontal line runs parallel to the x-axis, so its slope (change in y over change in x) is zero. Therefore, the y-value does not change as the x-value changes, and it is equal to a constant value (b).

For example, the equation of a horizontal line that passes through the point $(3,4)$ would be $y = 4$.

On the other hand, the equation of a vertical line is of the form $x = a$, where a is a constant. This is because a vertical line runs parallel to the y-axis, so its slope is undefined. Therefore, the x-value does not change as the y-value changes, and it is equal to a constant value (a).

For example, the equation of a vertical line that passes through the point $(3,4)$ would be $x = 3$.

- The equation of a horizontal line that passes through the point $(2,5)$ would be $y = 5$. This is because the line is parallel to the x-axis, and its slope (change in y over change in x) is zero. Therefore, the y-value does not change as the x-value changes, and it is equal to 5 for all points on the line.

- The equation of a vertical line that passes through the point $(-4,3)$ would be $x = -4$. This is because the line is parallel to the y-axis, and its slope is undefined. Therefore, the x-value does not change as the y-value changes, and it is equal to -4 for all points on the line.

It's worth noting that a horizontal line has a slope of zero, and the x-intercept is undefined, while a vertical line has an undefined slope, and the y-intercept is undefined.

In both cases, the line's equation is in the form of a single variable equals a constant.

Let's review an example:

Write an equation for the horizontal line that passes through (1,3).

Since the line is horizontal, the equation of the line is in the form of:

$y = b$,

where y always takes the same value of 3.

Thus, the equation of the line is:

$y = 3$.

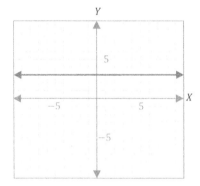

Once more example:

Write an equation for the vertical line that passes through (−5,5).

Since the line is vertical, the equation of the line is in the form of:

$x = a$,

where x always takes the same value of −5.

Thus, the equation of the line is: $x = -5$.

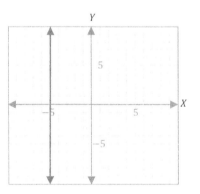

✍ Find the equation of each line.

30) Horizontal line: _____

   Passes through: (1,0)

31) Horizontal line: _____

   Passes through: (0,1)

32) Vertical line: _____

   Passes through: (7,3)

33) Horizontal line: _____

   Passes through: (0,0)

Graph a Horizontal or Vertical Line

A horizontal line is a line that runs along the x-axis, and its slope is equal to zero. To graph a horizontal line, we need to know the y-coordinate of the line, which is also known as the y-intercept.

For example, if we are given the equation of a horizontal line as $y = 3$, we know that the y-intercept is 3, which means that the line passes through the point (0,3) on the coordinate plane. To graph the line, we can simply plot that point and then extend the line horizontally to the right and to the left.

On the other hand, a vertical line is a line that runs along the y-axis. It has an undefined slope, and its x-coordinate is always a constant. To graph a vertical line, we need to know the x-coordinate, which is also known as the x-intercept.

For example, if we are given the equation of a vertical line as $x = 4$, we know that the x-intercept is 4, which means that the line passes through the point (4,0) on the coordinate plane. To graph the line, we can simply plot that point and then extend the line vertically upward and downward.

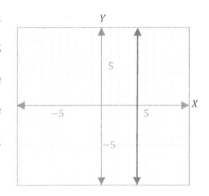

To graph the equation $y = -5$, we can first identify that this is a horizontal line, as the y-value is constant and equal to -5. We can then find the y-intercept, which is the point where the line crosses the y-axis. In this case, the y-intercept is $(0, -5)$. Then we can plot this point on the coordinate plane and extend the line horizontally, connecting all the points that have the y-value of -5. This line will be parallel to the x-axis and at a distance of 5 units below the x-axis.

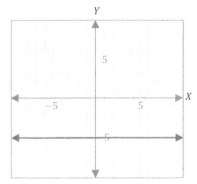

bit.ly/3D61ORj

Linear Functions

Another example:

Graph this equation: $x = 3$.

$x = 3$ is a vertical line and this equation tells you that every x-value is 3. You can consider some points that have an x-value of 3, then draw a line that connects the points.

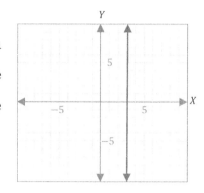

Graph this equation: $x = -2$.

$x = -2$ is a vertical line and this equation tells you that every x-value is -2. You can consider some points that have an x-value of -2, then draw a line that connects the points.

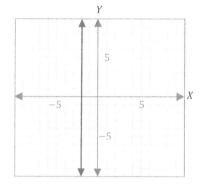

✎ Sketch the graph of each line.

34) Vertical line that passes through (2,6)

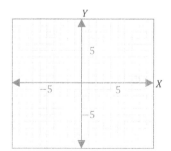

35) Horizontal line that passes through (5,3)

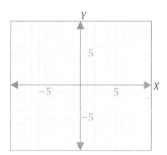

Graph an Equation in Point-Slope Form

To graph an equation in point-slope form, we first need to convert it to slope-intercept form $y = mx + b$. To do this, we can start by isolating y on one side of the equation. The point-slope form is $y - y_1 = m(x - x_1)$, we can add y_1 to both sides and we get $y = m(x - x_1) + y_1$.

For example, if we have an equation in point-slope form $y - 3 = -2(x - 4)$, we can convert it to slope-intercept form by adding 3 to both sides:

$y = -2x + 11$

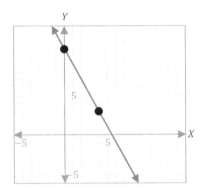

Now that we have the equation in slope-intercept form, we can graph it by finding the y-intercept, which is the point where the line crosses the y-axis. In this case, the y-intercept is $(0,11)$. We can then plot this point on the coordinate plane and use the slope of the line (-2) to find another point on the line. We can use the point $(x_1, y_1) = (4,3)$ that we had in the point-slope form as another point on the line.

Finally, we can connect the two points with a straight line, and the graph of the equation will be the line that passes through those points.

Another example:

To graph the line $y - 3 = -\frac{1}{3}(x + 5)$, first convert it to point-slope form by writing it as:

$y - 3 = -\frac{1}{3}(x + 5)$

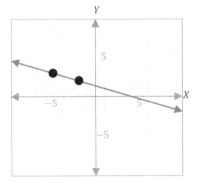

Then, use a point on the line, such as $(-5,3)$, and the slope of $-\frac{1}{3}$ to graph it.

Linear Functions

Plot the point $(-5,3)$ on the coordinate plane, then move down 1 unit and right 3 units to find the second point on the line $(-2,2)$. Connect these two points to graph the line.

One more example:

To graph the line $y - 2 = \frac{1}{2}(x + 4)$, first convert it to point-slope form by writing it as:

$$y - 2 = \frac{1}{2}(x + 4)$$

Then, use a point on the line, such as $(-6,1)$, and the slope of $\frac{1}{2}$ to graph it. Plot the point $(-6,1)$ on the coordinate plane, then move up 1 unit and right 2 units to find the second point on the line $(-4,2)$. Connect these two points to graph the line.

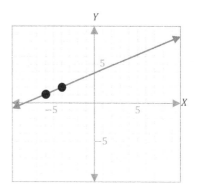

✎ Graph each equation.

36) $y + 3 = -\frac{1}{2}(x - 8)$

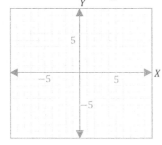

37) $y - 1 = 3(x + 6)$

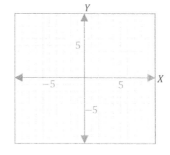

38) $y = (x + 11)$

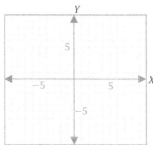

39) $y - 8 = -2(x - 1)$

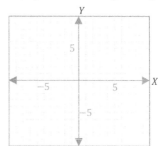

Equation of Parallel and Perpendicular Lines

The equation of a line in slope-intercept form is $y = mx + b$, where m is the slope of the line and b is the $y-$intercept. To find the equation of a line that is parallel to a given line, you can use the same slope and a different $y-$intercept. For example, if the equation of a given line is $y = 2x + 3$, a line that is parallel to it would have the equation $y = 2x + c$, where c is any real number.

To find the equation of a line that is perpendicular to a given line, you can use the negative reciprocal of the slope and the same $y-$intercept. For example, if the equation of a given line is $y = 2x + 3$, a line that is perpendicular to it would have the equation $y = -\frac{1}{2}x + 3$.

Alternatively, if a line has the equation $y = mx + b$, another line that is parallel to it will have the equation $y = mx + c$ where c is any real number. A line that is perpendicular to it will have the equation $y = -\frac{1}{m}x + d$ where d is any real number.

Let's review an example:

Find equation of a line parallel to $y = 6x + 4$ and passes though the point (0,7).

To find the equation of a line that is parallel to $y = 6x + 4$ and passes through the point (0,7), we can use the point-slope form of a linear equation. The point-slope form is $y - y_1 = m(x - x_1)$, where (x_1, y_1) is a point on the line, and m is the slope of the line. First, we know that the line is parallel to $y = 6x + 4$, which means that the slope of the line is 6.

Next, we can use the point (0,7) to find the $y-$intercept of the line. Substituting the point and the slope into the point-slope form gives: $y - 7 = 6(x - 0)$. Simplifying this equation gives: $y = 6x + 7$. So, the equation of the line that is parallel to $y = 6x + 4$ and passes through the point (0,7) is $y = 6x + 7$.

One more example:

Perpendicular to $y = -4x + 10$ and passes through the point (7,2).

The slope of $y = -4x + 10$ is -4. The negative reciprocal of that slope is: $m = \frac{(-1)}{(-4)} = \frac{1}{4}$. So, the perpendicular line has a slope of $\frac{1}{4}$.

Then: $y - y_1 = \frac{1}{4}(x - x_1)$ and now put in the point (7,2): $y - 2 = \frac{1}{4}(x - 7)$. Slope intercept $y = mx + b$ form: $y - 2 = \frac{1}{4}x - \frac{7}{4} \to y = \frac{1}{4}x + \frac{1}{4}$.

Another example:

Find the equation of a line perpendicular to $y = -2x + 7$ and passes through the point (1,3).

To find the equation of a line that is perpendicular to $y = -2x + 7$ and passes through the point (1,3), we can use the point-slope form of a linear equation and the negative reciprocal of the slope of the given line.

First, we know that the line is perpendicular to $y = -2x + 7$, which means that the slope of the line is the negative reciprocal of -2, which is $\frac{1}{2}$.

Next, we can use the point (1,3) to find the y-intercept of the line. Substituting the point and the slope into the point-slope form gives: $y - 3 = \frac{1}{2}(x - 1)$.

Simplifying this equation gives: $y = \frac{1}{2}x + 2.5$.

So, the equation of the line that is perpendicular to $y = -2x + 7$ and passes through the point (1,3) is $y = \frac{1}{2}x + 2.5$.

✎ Find the equation of each following.

40) Parallel to $y = -6x + 5$

　　Passes through (4,4)

41) perpendicular to $y = -\frac{1}{2}x - 4$

　　Passes through (7,1)

Compare Linear Function's Graph and Equations

Linear functions are functions that have a constant rate of change, represented by their slope. The slope of a linear function can be positive, negative, zero, or undefined. This slope is represented by the coefficient of the x −term in the equation of the line. The y −intercept of a linear function can be positive, negative, or zero. The y −intercept is represented by the constant term of the equation of the line.

The graph of a linear function is a straight line, and the slope of the line represents the rate of change of the function. A positive slope means that the function increases as x increases, while a negative slope means that the function decreases as x increases. A slope of zero means that the function does not change as x increases.

Two linear functions can have the same slope but different y −intercepts, which means that the two lines are parallel. They will have the same direction but different starting points. Two linear functions can have different slopes and different y −intercepts, which means that the two lines are not parallel. They will have different directions and different starting points.

To summarize, the slope of a linear function represents the rate of change of the function, and the y −intercept represents the starting point of the function. Two linear functions with the same slope but different y −intercepts are parallel, while two linear functions with different slopes are not parallel.

Let's review an example: Here's an example of comparing the equations of two linear functions.

Function 1: $y = 2x + 1$ and Function 2: $y = 2x - 3$

Both function 1 and function 2 have the same slope, which is 2. This means that the rate of change of both functions is the same, and the graph of both functions will have the same direction. However, function 1 has a y −intercept of 1, while function 2 has a y-intercept of -3. This means that the starting point of the two functions is different. The graph of function 1 will start at the point $(0,1)$ and the graph of function 2 will start at the point $(0,-3)$.

The graphs of both lines will be parallel to each other but will be shifted vertically. The graph of function 1 will be above the graph of function 2.

It's important to note that even though the two lines have different y-intercepts, both lines pass through an infinite number of points, so they are not different lines but just different representations of the same line.

Another example: Compare the following linear functions with each other.

Function 1: $x + 3y = 3$ and Function 2: $y = 3x - 2$

To compare, we evaluate the rate of change and y-intercept of two functions. First, we rearrange the given equation and write it in terms of y. So, we have: $y = -\frac{1}{3}x + 3$. We set the value of x to zero to get the y-intercept. Then: $x = 0 \rightarrow 0 + 3y = 9 \rightarrow y = 3$. The y-intercept is $(0,3)$.

Similarly for function 2, the y-intercept is $(0,-2)$, and the slope of the function is 3.

Now, you can see that the slopes are the negative reciprocal of together. This means that two lines are perpendicular.

✎ Compare the slope of the function A and function B.

42)

Function A: $y = 6x - 3$

Two-variable Linear Equations Word Problems

A two-variable linear equation is a type of equation that involves two variables, usually represented as x and y, with an exponent of 1. The solutions of a two-variable linear equation can be represented as ordered pairs, such as (x, y). These equations can take on several forms, including standard form, point-slope form, and intercept form.

The standard form of a two-variable linear equation is written as $ax + by = c$, where a, b, and c are real numbers and x and y are the variables.

The slope-intercept form of a two-variable linear equation is written as $y = mx + b$, where m represents the slope of the line and b represents the y-intercept.

The point-slope form of a two-variable linear equation is written as $y - y_1 = m(x - x_1)$, where (x, y) is a point on the line and m represents the slope of the line.

Solving two-variable linear equations word problems typically involves the following steps:

1. Read and understand the problem: Carefully read the problem and identify what information is given and what is being asked for.

2. Translate the problem into mathematical terms: Use mathematical symbols and equations to represent the information given in the problem.

3. Solve the equation: Use algebraic techniques to solve for the unknown variable.

4. Check your solution: Substitute your solution back into the original equation to make sure it is a valid solution.

5. Interpret the solution: Translate the solution back into the context of the problem and make sure it makes sense.

Here's an example of how to use these steps to solve a word problem:

Problem: A store sells two types of shirts, regular and sale shirts. The store sold a total of 50 shirts and earned $150 in revenue. Regular shirts sell for $10 each and sale shirts sell for $5 each. How many of each type of shirt did the store sell?

1. Read and understand the problem: The problem is asking how many regular and sale shirts the store sold.

2. Translate the problem into mathematical terms: Let x be the number of regular shirts and y be the number of sale shirts. We know that $x + y = 50$ and $10x + 5y = 350$.

3. Solve the equation: We can use either substitution or elimination method to solve for x and y. Using the first equation, we can find that $y = 50 - x$. Substituting this into the second equation we get: $10x + 5(50 - x) = 350$.

4. Check your solution: Once we solve the equation, we get $x = 20$ and $y = 30$, check this solution by plugging it back into the first equation and the second equation.

5. Interpret the solution: The store sold 20 regular shirts and 30 sale shirts.

It's also important to note that word problems could have more than one solution, no solution, or infinitely many solutions based on the given information, so it's important to check whether the solution obtained makes sense in the context of the problem.

✎ Answer the following questions.

43) John has an automated hummingbird feeder. He fills it to capacity, 8 fluid ounces. It releases 1 fluid ounce of nectar every day. Write an equation that shows how the number of fluid ounces of nectar left, y, depends on the number of days John has filled it, x.

44) The entrance fee to Park City is $9. Additionally, skate rentals cost $4 per hour. Write an equation that shows how the total cost, y, depends on the length of the rental in hours, x.

Chapter 4: Answers

1) 1
2) 2
3) −5
4) −2
5) $-\frac{1}{4}$
6) −6

7) $y = 4x + 16$
8) $y = 3x + 2$
9) $y = -5x - 3$
10) $y = -4x - 22$
11) $y = -3x - 15$

12)
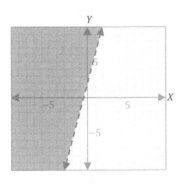

13)
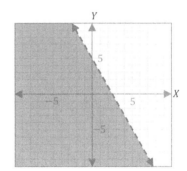

14) y −intercept: −1
   x −intercept: −3
   Slope: $-\frac{1}{3}$

15) y −intercept: 10
   x −intercept: −2
   Slope: 5

16) $y = -\frac{3}{2}x + 4$

17) $y = \frac{5}{3}x$

18) Point-slope form: $y + 6 = -2(x - 6)$
    Slope-intercept form: $y = -2x + 6$

19) Point-slope form: $y - 7 = 4(x + 7)$
    Slope-intercept form: $y = 4x + 35$

20) $y = -(x - 1)$
21) $y + 4 = 3x$
22) $y = \frac{2}{3}(x + 3)$
23) $y - 3 = -2(x + 4)$

24) $-\frac{6}{7}$
25) $-\frac{1}{2}$
26) 2
27) −6

28)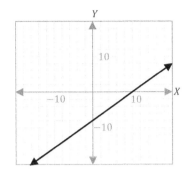

29)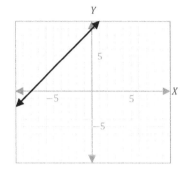

30) $y = 0$

31) $y = 1$

32) $x = 7$

33) $y = 0$

34)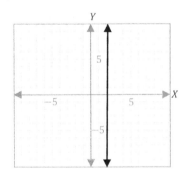

35)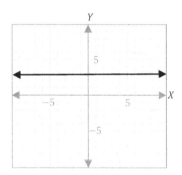

36)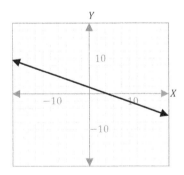

37)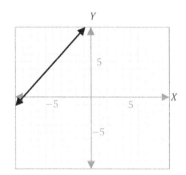

38)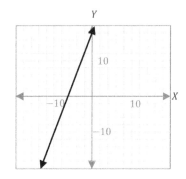

39)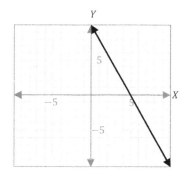

40) $y = -6x + 28$

41) $y = 2x - 13$

42) The absolute value of the slope of function A is 6 and it is steeper than the absolute value of the slope of function B (-1) and the slope of the two functions have different signs.

43) $y = -x + 8$

44) $y = 4x + 9$

CHAPTER 5: Inequalities and System of Equations

Math topics that you'll learn in this chapter:

- ☑ One–Step Inequalities
- ☑ Multi–Step Inequalities
- ☑ Compound Inequalities
- ☑ Write a Linear Inequality from a Graph
- ☑ Graph Solutions to One-step and Two-step Linear Inequalities
- ☑ Solve Advanced Linear Inequalities in Two-Variables
- ☑ Graph Solutions to Advance Linear Inequalities
- ☑ Absolute Value Inequalities
- ☑ Systems of Equations
- ☑ Find the Number of Solutions to a Linear Equation
- ☑ Write a System of Equations Given a Graph
- ☑ Systems of Equations Word Problems Top of Form
- ☑ Solve One-step and Two-step Linear Equations Word Problems
- ☑ Systems of Linear Inequalities
- ☑ Write Two-variable Inequalities in the Form of Word Problems

One−Step Inequalities

One-step inequalities are inequalities that involve a single operation, such as addition, subtraction, multiplication, or division. These inequalities can be solved by isolating the variable on one side of the inequality.

Inequality signs are: "less than" <, "greater than" >, "less than or equal to" ≤, and "greater than or equal to" ≥.

When solving one-step inequalities, remember to keep the inequality symbol consistent throughout the solution process. This means that when you multiply or divide both sides of an inequality by a positive number, the inequality symbol stays the same. However, when you multiply or divide both sides of an inequality by a negative number, the inequality symbol must be flipped. For example, if you have the inequality $-\frac{x}{2} > 4$, you would solve it as follows: $-\frac{x}{2} > 4 \to x < -8$.

Let's review an example:

Solve the inequality: $x + 7 \geq 11$.

To solve the inequality $x + 7 \geq 11$ for x, we need to isolate x on one side of the inequality. To do this, we can subtract 7 from both sides: $x + 7 \geq 11 \to x \geq 4$.

So, the solution is $x \geq 4$, which means that x can take on any value greater than or equal to 4. This solution can also be represented in interval notation as $x \in [4, \infty)$, which means all real numbers greater than or equal to 4.

One more example:

Solve the inequality: $x - 4 > -7$.

To solve the inequality $x - 4 > -7$ for x, we need to isolate x on one side of the inequality. To do this, we can add 4 to both sides: $x - 4 > -7 \to x > -3$.

So, the solution is $x > -3$, which means that x can take on any value greater than −3.

Inequalities and System of Equations

Another example:

Solve the inequality: $x + 3 \leq 2$.

The inverse (opposite) operation of addition is subtraction. In this inequality, 3 is added to x. To isolate x, we need to subtract 3 from both sides of the inequality. Then: $x + 3 \leq 2 \rightarrow x + 3 - 3 \leq 2 - 3 \rightarrow x \leq -1$.

The solution is: $x \leq -1$.

To better explain the problem, you can draw your attention to these examples:

Solve the inequality: $x - 1 > -7$.

1 is subtracted from x. Add 1 to both sides:

$x - 1 > -7 \rightarrow x - 1 + 1 > -7 + 1 \rightarrow x > -6$

Solve: $3x > -15$.

3 is multiplied to x. Divide both sides by 3. Then:

$3x > -15 \rightarrow \frac{3x}{3} > \frac{-15}{3} \rightarrow x > -5$

Solve: $-4x \leq 10$.

-4 is multiplied by x. Divide both sides by -4. Remember when dividing or multiplying both sides of an inequality by negative numbers, flip the direction of the inequality sign. Then:

$-4x \leq 10 \rightarrow \frac{-4x}{-4} \geq \frac{10}{-4} \rightarrow x \geq -2.5$

✎ Solve the following inequalities.

1) $y - 11 \leq 2$
2) $4x < -8$
3) $2a > 3$
4) $x - \frac{2}{3} \geq -2$
5) $3 + x \geq 7$
6) $-\frac{1}{2}x < -\frac{3}{2}$
7) $b < -2b + 3$
8) $-x \leq 5$

Multi–Step Inequalities

Solving multi-step inequalities is a process that involves several steps, similar to solving multi-step equations. The steps to solving multi-step inequalities are as follows:

1. Isolate the variable on one side of the inequality. This is done by using inverse operations such as addition, subtraction, multiplication, and division.
2. Simplify the inequality by combining like terms, if necessary.
3. Divide or multiply both sides by the same number, if the coefficient of the variable is not 1.
4. Change the direction of the inequality sign, if necessary, when dividing or multiplying both sides by a negative number.
5. Check your solutions by plugging them back into the original inequality and making sure they make the inequality true.
6. Graph the solution on a number line.

For example, consider the inequality: $2x + 4 > 10$.

1. Subtract 4 from both sides: $2x > 6$.
2. Divide both sides by 2: $x > 3$.
3. The solution to the inequality is $x > 3$. To check, we can substitute $x = 4$ into the original inequality and see that it does make the inequality true, $2 \times 4 + 4 > 10$.

When solving an inequality, it is important to remember that multiplying or dividing both sides of an inequality by a negative number changes the direction of the inequality.

Let's review another example:

Solve: $-5x + 12 \leq -2$.

To solve the inequality $-5x + 12 \leq -2$, we will follow these steps:

1. Isolate the variable on one side of the inequality by subtracting 12 from both sides: $-5x \leq -14$.
2. Divide both sides by -5: $x \geq 2.8$.
3. Reverse the inequality sign, as we're dividing by a negative number.
4. Check the solution by plugging it back into the original inequality and making sure it makes the inequality true: $-5(7) + 12 \leq -2 \rightarrow -35 + 12 \leq -2$, which is true.

The solution to this inequality is $x \geq 2.8$.

More examples to better explain:

Solve this inequality. $7x - 3 \geq 11$

In this inequality, 3 is subtracted from $7x$. The inverse of subtraction is addition.

Add 3 to both sides of the inequality: $7x - 3 + 3 \geq 11 + 3 \rightarrow 7x \geq 14$.

Now, divide both sides by 7. Then: $7x \geq 14 \rightarrow x \geq 2$.

The solution of this inequality is $x \geq 2$.

Solve this inequality. $5x - 2 \leq 13$

First, add 2 to both sides: $5x - 2 + 2 \leq 13 + 2$. Then simplify:

$$5x - 2 + 2 \leq 13 + 2 \rightarrow 5x \leq 15.$$

Now divide both sides by 5: $\frac{5x}{5} \leq \frac{15}{5} \rightarrow x \leq 3$.

Solve this inequality. $-2x + 5 < 3$

First, subtract 5 from both sides: $-2x + 5 - 5 < 3 - 5 \rightarrow -2x < -2$.

Divide both sides by -2. Remember that you need to flip the direction of the inequality sign. $-5x < 5 \rightarrow x > 1$.

✎ Solve the following inequalities.

9) $3x + 8 \leq -1$

10) $3 - x \geq 7$

11) $\frac{7-2x}{-3} > -1$

12) $-3 + 5x > 2$

13) $7 < -2x - 1$

14) $\frac{x+2}{5} \leq 1$

Compound Inequalities

Solving compound inequalities is a process that involves breaking down a compound inequality into a combination of two or more simple inequalities, and then solving each one separately.

Here are the steps to solve compound inequalities:

1. Recognize the type of compound inequality you are dealing with. Compound inequalities can be written in two forms: "and" or "or".
2. Break down the compound inequality into two simple inequalities. For example, if the compound inequality is of the form "and", then it will be in the form of "$x < a$ and $x > b$", which can be broken down into two simple inequalities: "$x < a$" and "$x > b$".
3. Solve each of the simple inequalities separately and find the solution set for each one.
4. Combine the solution sets of each of the simple inequalities to get the final solution set for the compound inequality.
5. Graph the solution set on the number line.

For example, consider the compound inequality $3x + 2 < 7$ or $2x - 5 \geq 0$.

1. Break down the compound inequality into two simple inequalities: $3x + 2 < 7$ and $2x - 5 \geq 0$.
2. Solve the first inequality: $3x + 2 < 7 \rightarrow 3x < 5 \rightarrow x < \frac{5}{3}$.
3. Solve the second inequality: $2x - 5 \geq 0 \rightarrow 2x \geq 5 \rightarrow x \geq \frac{5}{2}$.
4. The solution set for the compound inequality is $x < \frac{5}{3}$ or $x \geq \frac{5}{2}$.
5. We can graph the solution set on the number line.

Another example: Solve. $12 < -3x \leq 24$

To solve the compound inequality $12 < -3x \leq 24$, we will follow these steps:

1. Break down the compound inequality into two simple inequalities, $-3x > 12$ and $-3x \leq 24$.
2. Solve the first inequality: $-3x > 12 \rightarrow x < -4$.
3. Solve the second inequality: $-3x \leq 24 \rightarrow x \geq -8$.

Combine the solution sets of each of the simple inequalities to get the final solution set for the compound inequality: $-8 \leq x < -4$.

Another examples:

Solve. $6 < 3x \leq 24$

To solve this inequality, divide all sides of the inequality by 3. This simplifies the inequality as follows: $2 < x \leq 8$ or $(2,8]$.

Solve. $x - 5 < -9$ or $\frac{x}{5} > 3$.

Solve each inequality by isolating the variable:

$$x - 5 < -9 \rightarrow x - 5 + 5 < -9 + 5 \rightarrow x < -4.$$

Therefore: $\frac{x}{5} > 3 \rightarrow \frac{x}{5} \times 5 > 3 \times 5 \rightarrow x > 15$.

The solution to these two inequalities is:

$$x < -4 \text{ or } x > 15 \text{ or } (-\infty, -4) \cup (15, \infty).$$

✍ Solve the following inequalities.

15) $-4 < 2x < 5$

16) $0 < x - 3 \leq 1$

17) $3 \leq 1 - 2x \leq 5$

18) $\frac{3}{4}x < 2$ or $-x \geq -\frac{1}{3}$

19) $1 < \frac{2x-1}{3} < 2$

20) $-10 < -\frac{2}{5}a < 0$

21) $5x \leq 45$ and $x - 11 > -21$

22) $5x + 1 < 4$ or $-\frac{1}{2}x + 2 > -\frac{3}{2}$

Write a Linear Inequality from a Graph

When comparing two linear expressions using inequality symbols such as <, >, ≤, or ≥, they create linear inequalities. The graph of a linear inequality is represented by either a dashed or solid line, and one side is shaded. By analyzing the graph of a linear inequality and considering linear relationships, the equation of the linear inequality can be determined.

To write an equation from a graph of a linear inequality, these steps should be followed:

1. Determine if the inequality line is represented by a dashed or solid line. If it's a dashed line, the inequality symbol is < or >. If it's a solid line, the inequality symbol is ≤ or ≥.
2. Identify two points on the inequality line. These points can be used to determine the equation of the inequality.
3. Find the slope of the inequality line by using the two points identified in step 2. The slope formula is $m = \frac{y_2 - y_1}{x_2 - x_1}$.
4. Use the slope, a point on the line, and the y-intercept formula $y = mx + b$, where m is the slope of the line, (x, y) is a point on the line, and b is the $y-$intercept.
5. Observe the shaded part of the graph and determine if y is less than or greater than the obtained equation. You can use a point from the shaded part to find the sign of inequality.

Let's review an example: Write the slope-intercept form equation of the following graph.

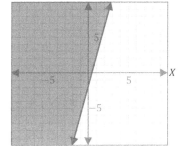

To write the inequality equation in slope-intercept form, you should find the $y-$intercept (b), and the slope (m), of the solid line. The value of b is -1 because the solid line passes through the y-axis at $(0, -1)$. Now, consider 2 points on the solid line to find the slope. You can use $(0, -1)$ and $(1, 2)$: $m = \frac{y_2 - y_1}{x_2 - x_1} = \frac{2-(-1)}{1-0} = 3 \rightarrow m = 3$.

Now, use the value of b and m and put them into the slope-intercept form formula: $y = mx + b \rightarrow y = 3x - 1$.

Determine the symbol of inequality: you have a solid line and the shaded part is above the line. So, the equation of the inequality is as follows: $y \geq 3x - 1$.

To better explain the problem, you can draw your attention to this example:

Write an equation from the following graph of the linear inequality.

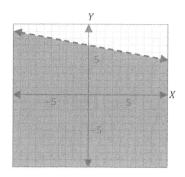

First, find the equation of the dashed line. For this purpose, by using two arbitrary points such as $(-5, 8)$ and $(0, 7)$ and the formula $m = \frac{y_2 - y_1}{x_2 - x_1}$, to evaluate the slope of the graph. We have: $m = \frac{7-8}{0-(-5)} = \frac{-1}{5}$. Then, by substituting in formula $y - y_0 = m(x - x_0)$, where (x_0, y_0) is a point like $(0, 7)$ on the line, the equation of the line is obtained: $y - 7 = \frac{-1}{5}(x - 0) \rightarrow y = \frac{-1}{5}x + 7$.

We put a point of the shaded area in the equation to determine the direction of the inequality. Considering point $(0,0)$, we have: $y = 0$ and $x = 0 \rightarrow \frac{-1}{5}(0) + 7 = 7$, which means that $y < \frac{-1}{5}x + 7$.

✎ Solve the following inequalities.

23) _____

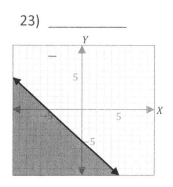

24) _____

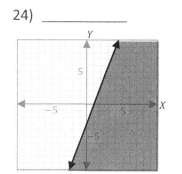

Graph Solutions to One-step and Two-step Linear Inequalities

Inequalities that you can solve using only one step are called one-step inequalities. Inequalities that you should take two steps to solve are called two-step inequalities.

To graph linear inequalities, use the following steps:

1. First, write the inequality in standard form. This means that the inequality should have the variable on one side and a constant on the other side.

    Example: $2x + 3 < 7$.

2. Next, solve for x by isolating x on one side of the inequality.

    Example: $2x < 4$.

3. Divide or multiply both sides of the inequality by the coefficient of x to get x by itself on one side. Example: $x < 2$.

4. Test a value that is less than the solution to see if it makes the inequality true and test a value that is greater than the solution to see if it makes the inequality false. Example: for $x < 2$, if we test $x = 1$, the inequality is true, and if we test $x = 3$, the inequality is false.

5. Graph the solution on a number line. If the inequality is < or >, use a dotted line to show that the solution is not included. If the inequality is ≤ or ≥, use a solid line to show that the solution is included. Example: $x < 2$ would be graphed as a dotted line with an open circle at 2.

6. Shade the appropriate side of the graph according to the inequality sign. < or > inequalities will be shaded on one side of the line, ≤ or ≥ will be shaded on both sides of the line. Example: $x < 2$ would be shaded to the left of the line.

For twostep inequalities:

1. Start by solving for x or y by isolating it on one side of the inequality.

2. Then solve for the next variable by substituting the first solution into the other inequality.

Inequalities and System of Equations

3. Test the solution by substituting it back into the original inequalities to see if it makes them true.
4. Graph the solution on a coordinate plane.
5. Shade the appropriate region on the coordinate plane according to the inequality sign.
6. Repeat the steps for any other solution.

Let's review an example: Solve the following inequality and graph the solution.
$$m - 2 \geq 3$$
Solve for m: $m - 2 \geq 3 \rightarrow m \geq 3 + 2 \rightarrow m \geq 5$. Now, graph $m \geq 5$. The inequality $m \geq 5$ means that m can be any number more than or equal to 5. m can be equal to 5, so you should use a filled-in circle located on 5. Also, m can be more than 5, so you should also draw an arrow pointing to the right:

Solve the following inequality and graph the solution: $-3q - 1 < 2$.

$-3q - 1 < 2$ is a two-step inequality. First, solve for q:
$$-3q - 1 < 2 \rightarrow -3q < 2 + 1 \rightarrow -3q < 3 \rightarrow q > -1.$$
Now, graph $q > -1$. The inequality $q > -1$ means that q can be any number more than -1. q can't be equal to -1, so you should use an open circle located on -1. Also, q can be more than -1, so you should also draw an arrow pointing to the right:

✍ Solve the following inequalities and graph the solution.

25) $x + 3 < 7$

26) $2x \geq -4$

27) $5 - x > 4$

28) $-1 < 2a - 7 \leq 3$

Solve Advanced Linear Inequalities in Two-Variables

The general form of a linear inequality in two-variable is $Ax + By < C$. To solve advanced linear inequalities in two variables, you can use the following steps:

1. Get the inequality in standard form, which means having the variables on one side and constants on the other, and the inequality symbol should be either $<, >, \leq,$ or $\geq$.
2. Isolate one of the variables in the inequality by performing the same operation on both sides of the inequality.
3. Plot the inequality on a coordinate plane, using the slope and y-intercept (if possible) of the equation to determine the correct half-plane.
4. Test a point that is not on the line to determine which side of the line to shade.
5. Graph the solution on the coordinate plane, shading the correct half-plane.
6. Write the solution using the inequality sign.
7. For example, consider the inequality: $2x + 3y > 6$.
8. To solve this inequality, we first want to isolate one of the variables on one side of the inequality sign. We can do this by subtracting $3y$ from both sides, resulting in: $2x > 6 - 3y$.
9. Next, we can divide both sides by 2, resulting in: $x > \frac{(6-3y)}{2}$.
10. Now we have the inequality in slope-intercept form, where x is the independent variable and $\frac{(6-3y)}{2}$ is the dependent variable. To graph this inequality, we can use a solid line if the inequality is "less than or equal to" or "greater than or equal to", and a dashed line if the inequality is "less than" or "greater than".
11. To find the solution set, we can pick a test point not on the line and

check whether it satisfies the inequality. For example, if we pick the point (0,0), we can substitute these values into the inequality and get: $0 > \frac{(6-3(0))}{2}$, which is not true. Therefore, the solution set is all the points that are above the line.

Let's review an example:

Solve the inequality $3x + 1 \geq 2y - x$.

First, convert to the general form. Subtract 1 from both sides of the inequality. So, $3x + 1 - 1 \geq 2y - x - 1 \rightarrow 3x \geq 2y - x - 1$.

Subtract $2y$ to both sides as $3x - 2y \geq 2y - x - 1 - 2y \rightarrow 3x - 2y \geq -x - 1$.

Also, add x to the sides: $3x - 2y + x \geq -x - 1 + x \rightarrow 4x - 2y \geq -1$.

The answer to this inequality is all ordered pairs in the form of (x, y), where $y \leq \frac{4x+1}{2}$. That is, $\{(x,y) | x \in \mathbb{R}, y \leq \frac{4x+1}{2}\}$.

Another example:

Solve the inequality $y - 5 < x - 3$.

Convert to the general form. Add 3 to both sides. So,

$$y - 5 + 3 < x - 3 + 3 \rightarrow y - 2 < x.$$

Now, subtract y from the sides $y - 2 < x \rightarrow x - y > -2$. The answer to this inequality is all ordered pairs in the form of (x, y), where $x > y - 2$. That is, $\{(x,y) | y \in R, x > y - 2\}$.

✍ Solve the following inequalities.

29) $1 + 2y < 3x + 4$

30) $y - 5x \geq 2y + 7$

31) $\frac{2x-3y}{2} > x - 1$

32) $y + 2 \leq 2x - 3$

33) $\frac{3}{2} > 3y - 2 + x$

34) $6x - 3y + 4 \leq 0$

Graph Solutions to Advance Linear Inequalities

To graph the solution to an advanced linear inequality in two variables, we can use the following steps:

1. Solve the inequality for one of the variables in terms of the other variable. This will put the inequality in slope-intercept form, where the independent variable is on one side of the inequality sign and the dependent variable is on the other side.

2. Determine the type of inequality. If the inequality is "less than or equal to" or "greater than or equal to", use a solid line to graph the equation. If the inequality is "less than" or "greater than", use a dashed line to graph the equation.

3. Plot the line on a coordinate plane, using the $x-$ and $y-$intercepts as additional points if necessary.

4. Pick a test point not on the line and substitute the coordinates into the inequality. If the inequality is true, the point is in the solution set. If the inequality is false, the point is not in the solution set.

5. Determine the solution set based on the inequality and the type of line used. If the inequality is "greater than" or "solid", the solution set is all the points above the line. If the inequality is "less than" or "dashed", the solution set is all the points below the line. If the inequality is "greater than or equal to" or "less than or equal to" include the line in the solution set.

6. Shade the solution set on the coordinate plane.

Keep in mind that the solution set to an inequality is a region in the coordinate plane and not a single point, this region is represented by shading the area of the coordinate plane that satisfies the inequality.

Here's an example:

Consider the inequality: $2x - 3y \leq 6$.

To solve this inequality, we can start by isolating one of the variables on one side of the inequality sign by adding $3y$ to both sides: $2x \leq 6 + 3y$.

Next, we can divide both sides by 2 to get: $x \leq \frac{(6+3y)}{2}$.

Now, we have the inequality in slope-intercept form, where x is the independent variable and $\frac{(6+3y)}{2}$ is the dependent variable. To graph this inequality, we can use a solid line, since the inequality is "less than or equal to".

To find the solution set, we can pick a test point not on the line and check whether it satisfies the inequality. For example, if we pick the point (1,1), we can substitute these values into the inequality and get: $1 \leq \frac{(6+3(1))}{2} = 4.5$, which is true. Therefore, the point (1,1) is in the solution set.

To graph this, we can plot the y-intercept which is $(0, -2)$ and the x-intercept which is $(3,0)$ and connect them using a dashed line. All the point below the line and on the line are the solution set.

Keep in mind that this is just one example, and the steps for solving and graphing advanced linear inequalities in two variables can vary depending on the specific inequality being solved.

✎ Solve the following inequalities.

35) $3x - 2y > 6$

36) $2y + x < 2x$

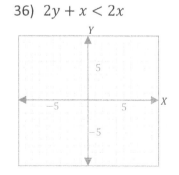

Absolute Value Inequalities

An absolute value inequality involves an absolute value expression, denoted by "| |", and a sign of inequality such as $<, >, \leq,$ or $\geq$. To solve an absolute value inequality, the first step is to transform it into a simple inequality by isolating the absolute value expression on one side of the inequality sign. The method used to transform the absolute value inequality into a simple inequality depends on the direction of the inequality. Depending on the direction of the inequality, one of the following methods can be used:

- For an inequality of the form $|a| < b$, we can write two separate inequalities: $a < b$ and $-a < b$.
- For an inequality of the form $|a| > b$, we can write two separate inequalities: $a > b$ and $a < -b$.
- For an inequality of the form $|a| \leq b$, we can write two separate inequalities: $a \leq b$ and $-a \leq b$.
- For an inequality of the form $|a| \geq b$, we can write two separate inequalities: $a \geq b$ and $a \leq -b$. After the inequalities are separated, they can be solved individually and the solution can be represented on a number line.

To solve absolute value inequalities, we can use the following steps:

1. Isolate the absolute value expression on one side of the inequality sign.
2. Split the inequality into two separate inequalities by using the rule that the absolute value of a number is always non-negative.
3. Solve each inequality separately, considering that the absolute value is non-negative.
4. Graph the solution on a number line, remembering that the solution is all the numbers that make the inequality true.

    For example: consider the inequality $|x - 3| \geq 2$.
    - In this example, the expression absolute value is isolated.

- Now we have two inequalities: $x - 3 \geq 2$ or $x - 3 \leq -2$.
- Solving each inequality separately: $x \geq 5$ or $x \leq 1$.
- Graph the solution on a number line: $(-\infty, 1] \cup [5, +\infty)$.

Keep in mind that when solving absolute value inequalities, it is important to remember that the absolute value of a number is always non-negative, and to split the inequality into two separate inequalities. Also, the solution to an absolute value inequality is a set of numbers represented by an interval on the number line.

In other words, to solve the absolute value inequality depending on the direction of the inequality can be used one of the following methods:

- To solve x in the inequality $|ax + b| < c$, you must solve $-c < ax + b < c$.
- To solve x in the inequality $|ax + b| > c$ you must solve $ax + b > c$ and $ax + b < -c$.

For better explain, consider the following example:

Solve: $|2x + 1| < 7$.

Since the inequality sign is $<$, rewrite the inequality to: $-7 < 2x + 1 < 7$. Then, solve the inequality: $-7 - 1 < 2x + 1 - 1 < 7 - 1 \rightarrow -8 < 2x < 6$. Now, divide each section by 2: $-8 < 2x < 6 \rightarrow -4 < x < 3$.

You can also write this solution using the interval symbol: $(-4, 3)$.

🖉 Solve the following inequalities and graph the solution.

37) $|2 + x| < 1$

38) $\left|-\frac{2}{3}x\right| \geq 2$

39) $|3x - 1| \leq 3$

40) $|-2x + 3| > 1$

System of Equations

A system of equations is a set of two or more equations that contain multiple variables. To solve a system of equations, one common method is the elimination method. The elimination method uses the property of equality that allows you to add or subtract the same value from both sides of an equation.

For example, consider the system of equations: $x - y = 1$ and $x + y = 5$.

To use the elimination method, we can start by adding the same value to both sides of each equation. In this case, we can subtract x from both sides of the first equation and y from both sides of the first equation. Now, we have: $x - x - y = -x + 1 \rightarrow -y = -x + 1$, and $-y - y = -x - y + 1 \rightarrow -2y = -x - y + 1$. Next, you can see that there is the expression $x + y$ from the first side of the second equation on the left side of the obtained equation as follows:

$-2y = -x - y + 1 \rightarrow -2y = -(x + y) + 1$.

Now, substitute 5 instead of $x + y$: $-2y = -(5) + 1 \rightarrow -2y = -4 \rightarrow y = 2$.

Now we can substitute the value of y into the second equation:

$y = 2 \rightarrow x + 2 = 5 \rightarrow x = 3$.

So, the solution to this system of equations is $(x, y) = (3, 2)$.

Here's another example of a system of equations and how to solve it using the elimination method:

Consider the system of equations: $2x + y = 12$ and $4x + 6y = 24$.

1. To use the elimination method, we can start by multiplying one of the equations by a constant. In this case, we can multiply the first equation by 2. Now, we have: $4x + 2y = 24$ and $4x + 6y = 24$.

2. Next, we can subtract the first equation from the second equation. This will eliminate one of the variables and leave us with a simpler equation: $4x + 6y - (4x + 2y) = 24 - (24) \rightarrow 2y = 0 \rightarrow y = 0$.

3. Now, we can solve for y: $y = 0$.
4. Now that we know the value of y, we can substitute it back into one of the original equations to find the value of x. In this case, we can substitute $y = 0$ into the first equation: $2x + (0) = 12 \rightarrow 2x = 12 \rightarrow x = 6$.
5. So, the solution to this system of equations is $(x, y) = (6, 0)$.

Keep in mind that there are many methods for solving systems of equations, and the elimination method is just one of them. Depending on the specific system of equations, some methods may be more efficient or easier to use than others.

As another example:

What is the value of xy in this system of equations? $\begin{cases} -x + 3y = -2 \\ 3x + y = -2 \end{cases}$

Solving a System of Equations by Elimination: multiply the first equation by (3) then add it to the second equation.

$\begin{matrix} 3(-x + 3y = -2) \\ 3x + y = -2 \end{matrix} \rightarrow \begin{matrix} -3x + 9y = -6 \\ 3x + y = -2 \end{matrix} \rightarrow 3x + y + (-3x + 9y) = -2 + (-6)$

$\rightarrow 10y = -8 \rightarrow y = -\frac{4}{5}$

Plug in the value of y into one of the equations and solve for x.

$-x + 3\left(-\frac{4}{5}\right) = -2 \rightarrow -x + \left(-\frac{12}{5}\right) = -2 \rightarrow -x = -2 + \frac{12}{5} \rightarrow x = -\frac{2}{5}$.

Thus, $xy = -\frac{2}{5} \times \left(-\frac{4}{5}\right) = +\frac{8}{25} = 0.32$.

✎ Solve the following system of equations.

41) $2x - y = 3$ and $x + 2y = -1$

42) $\begin{cases} x + 2y = 4 \\ 3x - y = 6 \end{cases}$

43) $\begin{cases} 3x + y = 6 \\ x - y = -2 \end{cases}$

44) $4x + 3y = 12$ and $2x - y = 18$

45) $\begin{cases} 2a + b = 7 \\ 4a - b = 5 \end{cases}$

46) $\begin{cases} x - y = 4 \\ 2x - 2y = 10 \end{cases}$

Find the Number of Solutions to a Linear Equation

The linear equation is a kind of equation with the highest degree of 1. In other words, in a linear equation, there is no variable with an exponent more than 1. A linear equation's graph always is in the form of a straight line, it's called a "linear equation".

The linear equation has no solution if by solving a linear equation you get a false statement as an answer.

The linear equation has one solution if by solving a linear equation you get a true statement for a single value for the variable.

The equation has infinitely many solutions if by solving a linear equation you get a statement that is always true.

Let's review an example:

How many solutions does the following equation have?

$$7 - 4p = -4p$$

Solve for p: $7 - 4p = -4p \rightarrow 7 = -4p + 4p \rightarrow 7 = 0$.

$7 = 0$ is a false statement. The linear equation has no solution because by solving the linear equation you get a false statement as an answer.

For another example:

How many solutions does the following equation have?

$$12h = 3h + 27$$

Solve for h: $12h = 3h + 27 \rightarrow 12h - 3h = 27 \rightarrow 9h = 27 \rightarrow h = 3$.

$h = 3$ is a true statement for a single value for the variable. So, the linear equation has one solution because by solving a linear equation you get a true statement for a single value for the variable.

One more example:

How many solutions does the following equation have?

$$8n - 2n = 6n$$

Simplify the left side of the equation: $8n - 2n = 6n \to 6n = 6n$.

$6n = 6n$ is a statement that is always true. So, the equation has infinitely many solutions because by solving a linear equation you get a statement that is always true.

For more explanation about the problem, pay attention to the following example:

How many solutions does the following equation have?

$$42 - 4x = 2(x - 6)$$

Solve for x:

$$42 - 4x = 2(x - 6) \to 42 - 4x = 2x - 12$$
$$\to 42 - 4x - 42 = 2x - 12 - 42$$
$$\to -4x = 2x - 54$$
$$\to -4x - 2x = 2x - 54 - 2x$$
$$\to -6x = -54$$
$$\to x = 9$$

$x = 9$ is a true statement for a single value for the variable. So, the linear equation has one solution.

✎ How many solutions do the following equations have?

47) $2x + 3 = 14$

48) $5a = 9a - 6a$

49) $1 - c + 4 = -c - 3$

50) $2y + 5y = -3y$

51) $-x + 8 + 5x = 4x - 2$

52) $3x + 2 = x - 4$

Write a System of Equations Given a Graph

A system of equations is a group of two or more equations that share the same variables. In particular, a system of linear equations is a set of equations that can be satisfied by the same set of values for the variables.

To write a system of equations given a graph, it is important to understand that each line in the graph represents a linear equation. By determining the equation of each line, you can create a system of equations.

To write the equation of a line, you can use the slope-intercept form, $y = mx + b$. To find the slope (m) of a line, you can use two points on the line and plug them into the slope equation: $m = \frac{change\ in\ y}{change\ in\ x}$. To determine the y-intercept (b), you can see at which point the line crosses the y-axis. Then you can use the value of the slope and y-intercept to construct the line's equation in slope-intercept form. Repeat this process for the second line to get the second equation of the system of equations.

Let's review an example:

Write a system of equations for the following graph.

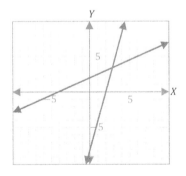

First, look at the red line. write its equation by identifying the slope (m) and y-intercept (b): to find the slope, you can use any two points on the line and plug in the slope equation: $m = \frac{change\ in\ y}{change\ in\ x}$. Here choose the points $(0,2)$ and $(-4,0)$: $m = \frac{0-2}{-4-0} = \frac{1}{2}$. The red line crosses the y-axis at $(0,2)$, so the y-intercept is $b = 2$. Now, write the equation of the red line: $y = \frac{1}{2}x + 2$.

You can find the blue line's equation in the same way. Use the points $(0,-8)$ and $(2,0)$: $m = \frac{0-(-8)}{2-0} = 4$. The blue line crosses the y-axis at $(0,-8)$, so the y-intercept is $b = -8$. Now, write the equation of the blue line:

$y = 4x - 8$. Therefore, the graph shows the following system of equations:
$$\begin{cases} y = \frac{1}{2}x + 2 \\ y = 4x - 8 \end{cases}.$$

For another example:

Write the system of equations of the following graph.

Consider two arbitrary points on each of the lines. Then calculate the slope of the lines using the formula: $m = \frac{y_2 - y_1}{x_2 - x_1}$, where (x_1, y_1) and (x_2, y_2) are two points on the line. Therefore,

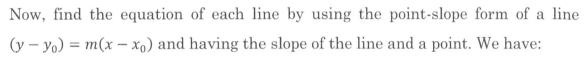

Red line: $(-1, 0)$ and $(0, -1) \rightarrow m = -1$

Blue line: $(-3, 0)$ and $(0, 5) \rightarrow m = \frac{5}{3}$

Now, find the equation of each line by using the point-slope form of a line $(y - y_0) = m(x - x_0)$ and having the slope of the line and a point. We have:

Red line: $m = -1$ and $(0, -1) \rightarrow y - (-1) = -1(x - 0) \rightarrow y + 1 = -x$

Blue line: $m = \frac{5}{3}$ and $(-3, 0) \rightarrow y - 0 = \frac{5}{3}(x - (-3)) \rightarrow y = \frac{5}{3}x + 5$

Finally, the system of linear equations shown in the above graph is as follows:
$$\begin{cases} y + 1 = -x \\ y = \frac{5}{3}x + 5 \end{cases}$$

✒ Solve the following system of equations.

53) _____

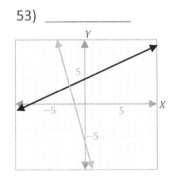

54) _____

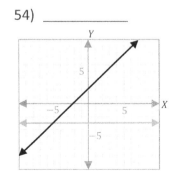

Systems of Equations Word Problems Top of Form

To solve systems of equations word problems, follow these steps:

1. Read the problem carefully and identify the variables and the information given.
2. Translate the information from the problem into equations. This will typically involve writing an equation for each piece of information given.
3. Write the system of equations using the equations you just created.
4. Solve the system of equations using one of the methods (substitution, elimination, or graphing)
5. Check your solution by substituting it back into the original equations and making sure it makes sense in the context of the problem.

Here is an example:

A company produces two products, x and y. Each unit of x costs $20 and each unit of y costs $30. The company's daily revenue is $700. How many units of x and y the company should produce to maximize the revenue?

1. Identify the variables: x (number of units of product x) and y (number of units of product y)
2. Translate the information into equations: $20x + 30y = 700$ (the revenue of the company)
3. Write the system of equations: $20x + 30y = 700$.
4. Solve the system of equations: You can use any method you prefer, in this case, I will use substitution: $x = \frac{700-30y}{20}$.
5. Check your solution: Now, you know that $x = \frac{700-30y}{20}$ and you can substitute this into the equation $20x + 30y = 700$, and check if the equation holds true.
6. Now you can use this information to find the values of x and y that maximize the revenue.

Note that in this example you can use the graph method and find the solution that way too, it's your choice. Also, sometimes the word problems will be a bit more complex, and you may need to use multiple equations or algebraic manipulations to solve them.

One more example:

Tickets to a movie cost $7 for adults and $4 for students. A group of friends purchased 24 tickets for $123. How many adult tickets did they buy?

Let x be the number of adult tickets and y be the number of student tickets. There are 24 tickets.

Then: $x + y = 24$. The cost of adults' tickets is $7 and for students, it is $4, and the total cost is $123. So, $7x + 4y = 123$.

Now, we have a system of equations: $\begin{cases} x + y = 24 \\ 7x + 4y = 123 \end{cases}$.

To solve this system of equations, multiply the first equation by -4 and add it to the second equation: $-4(x + y = 24) \rightarrow -4x - 4y = -96$. So, we have:

$7x + 4y = 123 + (-4x - 4y) = 123 - 96 \rightarrow 3x = 27 \rightarrow x = 9$.

Now, substitute the obtained value of x in one of the equations to get the value of y. Use the first equation. Therefore, $x = 9 \rightarrow 9 + y = 24 \rightarrow y = 15$.

There are 9 adult tickets and 15 student tickets. Now, check your answers by substituting solutions into the original equations.

$x + y = 24 \rightarrow 9 + 15 = 24$, $7x + 4y = 123 \rightarrow 7(9) + 4(15) = 123 \rightarrow 63 + 60 = 123$.

The solutions are correct in both equations.

✎ Solve the following word problem.

55) Sara and Jack are saving money. Sara starts with $40 and saves $25 per week. Jack starts with $110 and saves $15 per week. After how many weeks do they have the same amount of money?

Solve One-step and Two-step Linear Equations Word Problems

To solve one-step and two-step linear equations word problems, you should follow these steps:

1. Read the problem carefully and identify the variables and the information given.
2. Translate the information from the problem into an equation. This will typically involve writing an equation for each piece of information given.
3. Solve the equation using one of the methods (addition, subtraction, multiplication, or division)
4. Check your solution by substituting it back into the original equation and making sure it makes sense in the context of the problem.

Here is an example of a one-step linear equation word problem:

A store is selling a shirt for $20. If you apply a discount of 20%, how much will the shirt cost?

1. Identify the variables: x (cost of the shirt).
2. Translate the information into an equation: $x = 20$.
3. Solve the equation: $x = 20 \times \left(\frac{80}{100}\right) = 16$.
4. Check your solution: The shirt costs $16 after the discount.

Here is an example of a two-step linear equation word problem:

John has two times more money than Jane. If John has $120, how much money does Jane have?

1. Identify the variables: x (amount of money Jane has) and y (amount of money John has).
2. Translate the information into an equation: $y = 2x$.
3. Solve the equation:

$y = 2x$, and $y = 120 \rightarrow 120 = 2x \rightarrow x = \frac{120}{2} \rightarrow x = 60$.

4. Check your solution: If John has $120, then Jane has $60.

It's worth mentioning that it's important to pay attention to the units of measurement when solving word problems. Make sure to convert units of measurement if necessary, and always include the units in your final answer.

Also, when solving two-step linear equation word problems, you will often need to use inverse operations (addition, subtraction, multiplication, or division) to isolate the variable. Remember that to undo an operation, you should do the opposite operation.

In addition, be mindful of the signs of inequalities if the problem is asking for a range of solutions or solutions that are greater or lesser than a certain value.

One more example:

Larry is in a chocolate shop and is going to buy some chocolates for his friends. He chooses 7 chocolates with a flower design and in addition, he also chooses some packs of five chocolates. If the total number of chocolates he has bought is 32, write an equation that you can use to find c, the number of packs of three chocolates. How many packs of three chocolates has Larry bought?

Larry chooses some packs of three chocolates and n is the number of packs of three chocolates. We can write it as follows: $5c$. He also chooses 7 chocolates with a flower design. The total number of chocolates he has bought is 32. Therefore, we can complete the equation as follows: $5c + 7 = 32$. The equation $5c + 7 = 32$ can be used to find how many packs of three chocolates Larry bought. Solve the equation for c: $5c + 7 = 32 \rightarrow 5c = 32 - 7 \rightarrow 5c = 25 \rightarrow c = 25 \div 5 = 5 \rightarrow c = 5$. So, he bought 5 packs of five chocolates.

✎ Solve each word problem.

56) A rectangle is $3\,m$ tall and $5\,m$ wide. if its width is enlarged to $2\,m$ without changing its perimeter, then find the new length of the rectangle?

Systems of Linear Inequalities

A system of linear inequalities is a group of two or more linear inequalities that share one or more variables. The goal of solving a system of linear inequalities is to find the values of the variables that satisfy all of the inequalities in the system. To solve a system of linear inequalities, one can use a variety of methods, such as graphing or substitution.

Graphing is a common method used to solve systems of linear inequalities. This method involves plotting each inequality on a coordinate plane and finding the region where the solutions to all of the inequalities overlap. The solutions to the system of inequalities are the points that lie within the region where the solutions of the inequalities overlap.

Another method is Substitution, this is where we substitute the expression of one equation of the system in the other one, then solve for the variable and substitute back in the first equation and solve for the other variable.

In general, solving a system of linear inequalities can be a bit more complex than solving a single linear inequality, but with the right methods and strategies, it can be done. And it's important to understand that the solutions to a system of linear inequalities are not a single value, but a set of values (a range) that make the system true.

Let's review an example:

Solve the following system of inequalities:

$$\begin{cases} 8x - 4y \leq 12 \\ 3x + 6y \leq 12 \\ y \geq 0 \end{cases}$$

To solve this system of inequalities, we can use the method of graphing.

First, we'll graph the first inequality: $8x - 4y \leq 12$. We can start by putting it in slope-intercept form: $y = 2x - 3$. We can see that the inequality is pointing towards the direction of "less than or equal to", thus we'll shade the side of the line that does not include the line.

Second, we'll graph the second inequality: $3x + 6y \leq 12$. We can start by putting it in slope-intercept form: $y = -\frac{x}{2} + 2$. As the inequality is also pointing towards the direction of "less than or equal to", we'll shade the side of the line that does not include the line.

Finally, the third inequality is a non-strict inequality, $y \geq 0$, which means that **y** is greater than or equal to 0.

Now, we'll look at the graph, where the intersection of the shaded areas gives us the solution of the system of inequalities. And the answer is the solution is the area of intersection between the two shaded areas, which is the area above the line $y = 2x - 3$ and below the line $y = -\frac{x}{2} + 2$, and $y \geq 0$.

We can also represent the solution of the system by writing the ordered pair (x, y) that makes all the inequalities true.

The solution to the system is the set of all ordered pairs (x, y) that satisfies the system of inequalities. It's a set of points that form the shaded area on the graph.

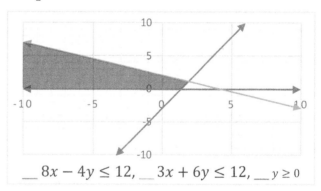

$8x - 4y \leq 12, \quad 3x + 6y \leq 12, \quad y \geq 0$

✍ Solve each following system of inequalities.

57) $\begin{cases} 3x - 2y \geq -1 \\ x + y < -2 \\ y > \frac{5}{3}x - 4 \end{cases}$

58) $\begin{cases} 3x + y \geq -3 \\ x - 2y < 2 \end{cases}$

59) $\begin{cases} y \leq 2 \\ 2x + y < -2 \\ x - y \leq -1 \end{cases}$

60) $\begin{cases} x + y > -3 \\ x - y \leq -2 \end{cases}$

Write Two-variable Inequalities in Form of Word Problems

The two-variable linear inequalities describe a not equal relationship between 2 algebraic statements which contain two different variables. A Two-variables linear inequality is created when two variables are involved in the equation and inequality symbols ($<$, $>$, $\leq$, or $\geq$) are used to connect 2 algebraic expressions.

The two-variables linear inequalities solution can be expressed as an ordered pair (x, y). When the x and y values of the ordered pair are replaced in the inequality they make a correct expression.

The method of solving a linear inequality is like the method of solving a linear equation. The difference between these two methods is in the symbol of inequality. You solve linear inequalities word problems in the same way as linear equations word problems:

Step 1: First, read the problem and try to write an inequality in words in a way that expresses the situation well.

Step 2: Represent each of the given information with numbers, symbols, and variables.

Step 3: Compare the inequality you've written in words with the given information that you've represented with numbers, symbols, and variables. Then rewrite what you've found out in form of an inequality expression.

Step 4: Simplify both sides of the inequality. Once you find the values, you have one of these inequalities:
- Strict inequalities: In this case, both sides of the inequalities can't be equal to each other.
- Non-strict inequalities: In this case, both sides of the inequalities can be equal.

Let's review an example:

Sara plans to hold a party. She plans to order pizza and pasta from a restaurant for dinner. The cost of each pizza is $45 and the cost of each pasta is $30. She hopes to spend no more than $400 on dinner. Write a linear inequality so that x represents the number of pizzas and y represents the number of pasta.

First, try to write an inequality in words in a way that expresses the situation well: The cost of the pizzas plus the cost of the pasta is no more than $400. Now, represent each of the given information with numbers, symbols, and variables: The cost of the pizzas is $45 times the number of pizzas, which is x. The product is $45x$. The cost of the pasta is $30 times the number of pasta, which is y. The product is $30y$. No more than means less than or equal ($\leq$). Rewrite what you've found out in form of an inequality expression: $45x + 30y \leq 400$.

✎ Solve.

61) A baker is buying barley flour and wheat flour and can spend no more than $840. Each kilogram of wheat flour costs $2.5, and each kilogram of barley flour costs $5.8.

    Write an inequality that represents all possible combinations of x, the amount of wheat flour, and y, the amount of barley flours the customer can buy.

62) An influencer has two different jobs. His combined work schedules consist of less than 48 hours in week.

    Write an inequality that shows the best solution set for all possible combinations of x, the number of hours she worked at her first job, and y, the number of hours she worked at her second job, in one week?

Chapter 5: Answers

1) $y \leq 13$
2) $x < -2$
3) $a > \frac{3}{2}$
4) $x \geq -\frac{4}{3}$
5) $x \geq 4$
6) $x > 3$
7) $b < 1$
8) $x \geq -5$
9) $x \leq -3$
10) $x \leq -4$
11) $x > 2$

12) $x > 1$
13) $x < -4$
14) $x \leq 3$
15) $(-2, \frac{5}{2})$
16) $(3, 4]$
17) $[-2, -1]$
18) $\left(-\infty, \frac{8}{3}\right]$
19) $\left(2, \frac{7}{2}\right)$
20) $(0, 25)$
21) $(-10, 9]$
22) $(-\infty, 7)$

23) $y \leq -x - 5$

24) $y \leq 2x - 2$

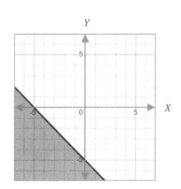

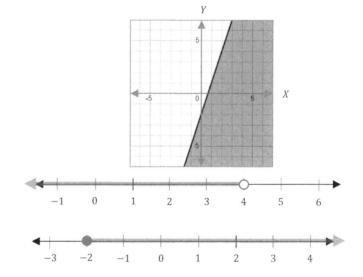

25) $x + 3 < 7$

26) $2x \geq -4$

27) $5 - x > 4$

28) $-1 < 2a - 7 \leq 3$

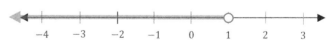

Inequalities and System of Equations

29) $\{(x,y) | x \in \mathbb{R}, y < \frac{3}{2}(x+1)\}$

30) $\{(x,y) | x \in \mathbb{R}, y \le -(5x+7)\}$

31) $\{(x,y) | x \in \mathbb{R}, y < \frac{2}{3}\}$

32) $\{(x,y) | x \in \mathbb{R}, y \le 2x - 5\}$

33) $\{(x,y) | y \in \mathbb{R}, x < \frac{7}{2} - 3y\}$

34) $\{(x,y) | y \in \mathbb{R}, x \le \frac{3y-4}{6}\}$

35) $3x - 2y > 6$

36) $2y + x < 2x$

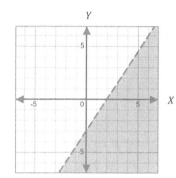 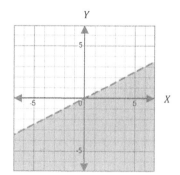

37) $-3 < x < -1$

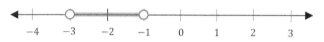

38) $x \le -3$ or $x \ge 3$

39) $-\frac{2}{3} \le x \le \frac{4}{3}$

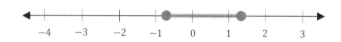

40) $x > 2$ or $x < 1$

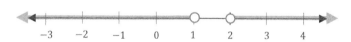

41) $x = 1$ and $y = -1$

42) $x = \frac{16}{7}$ and $y = \frac{6}{7}$

43) $x = 1$ and $y = 3$

44) $x = 6.6$ and $y = -4.8$

45) $a = 2$ and $b = 3$

46) No solution

47) One solution

48) One solution

49) No solution

50) One solution

51) No solution

52) One solution

Effortless Math Education

53) $\begin{cases} x - 2y = -8 \\ 9x + 2y = -12 \end{cases}$

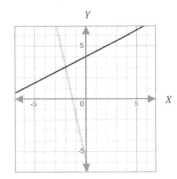

54) $\begin{cases} y = -3 \\ -x + y = 2 \end{cases}$

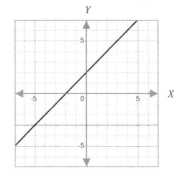

55) 7 weeks

56) $1\ m$

57)

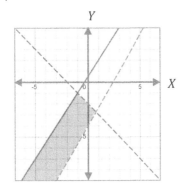

58)

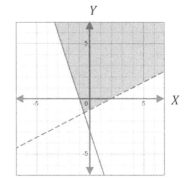

59)

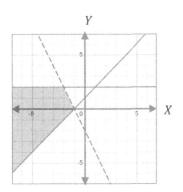

60)

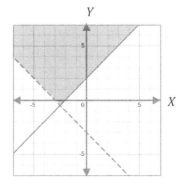

61) $2.5x + 5.8y \leq 840$

62) $x + y \leq 48$

CHAPTER 6: Quadratic

Math topics that you'll learn in this chapter:

- ☑ Solving Quadratic Equations
- ☑ Graphing Quadratic Functions
- ☑ Solve a Quadratic Equation by Factoring
- ☑ Transformations of Quadratic Functions
- ☑ Quadratic Formula and the Discriminant
- ☑ Characteristics of Quadratic Functions: Equations
- ☑ Characteristics of Quadratic Functions: Graphs
- ☑ Complete a Function Table: Quadratic Functions
- ☑ Domain and Range of Quadratic Functions: Equations
- ☑ Factor Quadratics: Special Cases
- ☑ Factor Quadratics Using Algebra Tiles
- ☑ Write a Quadratic Function from Its Vertex and Another Point

Solving Quadratic Equations

To solve a quadratic equation, first write the equation in the form of:

$$ax^2 + bx + c = 0,$$

where a, b, and c are coefficients and x is the variable. This is called the standard form of a quadratic equation.

Once you have the equation in this form, you can use the quadratic formula:

$$x = \frac{-b \pm \sqrt{b^2 - 4ac}}{2a}$$

To find the solutions (also known as roots or zeros) of the equation. If you can factor the equation, you can also set each factor equal to zero and solve for x.

For example, if the equation is $x^2 + 3x = 4$, you can write it in the standard form of a quadratic equation: $x^2 + 3x - 4 = 0$.

Then use the quadratic formula to find the solutions.

$$x = \frac{-b \pm \sqrt{b^2 - 4ac}}{2a}$$

Where $a = 1$, $b = 3$, and $c = -4$.

$$x = \frac{-3 \pm \sqrt{3^2 - 4(1)(-4)}}{2(1)} = \frac{-3 \pm \sqrt{9 + 16}}{2} = \frac{-3 \pm 5}{2}$$

So, the solutions are: $x = \frac{-3+5}{2} = 1$ and $x = \frac{-3-5}{2} = -4$.

The solutions to this equation are $x = 1$ and $x = -4$.

We can also use factoring to solve this equation. To solve the equation by factoring, we need to find two numbers that multiply by -4 and add to 3. These numbers are -1 and -4, so we can factor the equation as:

$$x^2 + 3x - 4 = (x - 1)(x + 4) = 0$$

So, we have $x - 1 = 0$ and $x + 4 = 0$.

So, the solutions to this equation are $x = 1$ and $x = -4$.

A few more examples:

Find the solutions of the following quadratic equation.

Quadratic

$$x^2 - x - 12 = 0.$$

Factor the quadratic by grouping. We need to find two numbers whose sum is -1 (from $-x$) and whose product is -12. Those numbers are 3 and -4. Then:

$$x^2 - x - 12 = 0 \to x^2 + 3x - 4x - 12 = 0 \to (x^2 + 3x) + (-4x - 12) = 0.$$

Now, find common factors: $(x^2 + 3x) = x(x + 3)$ and $(-4x - 12) = -4(x + 3)$. We have two expressions $(x^2 + 3x)$ and $(-4x - 12)$ and their common factor is $(x + 3)$. Then:

$$(x^2 + 3x) + (-4x - 12) = 0 \to x(x + 3) - 4(x + 3) = 0$$
$$\to (x + 3)(x - 4) = 0.$$

The product of two expressions is 0. Then:

$$(x + 3) = 0 \to x = -3 \text{ or } (x - 4) = 0 \to x = 4.$$

Solve the equation below:

$$x^2 + 4x + 3 = 0$$

Use quadratic formula: $x_{1,2} = \frac{-b \pm \sqrt{b^2 - 4ac}}{2a}$, $a = 1$, $b = 4$ and $c = 3$.

Then: $x_{1,2} = \frac{-4 \pm \sqrt{4^2 - 4 \times 1(3)}}{2(1)} = \frac{-4 \pm \sqrt{16 - 12}}{2} = \frac{-4 \pm 2}{2}$,

$x_1 = \frac{-4+2}{2} = -1$, $x_2 = \frac{-4-2}{2} = -3$.

Find the zeros of the quadratic:

$$x^2 + 9x + 8 = 0$$

Factor: $x^2 + 9x + 8 = 0 \to (x + 8)(x + 1) = 0 \to x = -8$, or $x = -1$.

✎ Solve each equation.

1) $x^2 - x - 2 = 0$

2) $x^2 - 6x + 8 = 0$

3) $x^2 - 4x + 3 = 0$

4) $x^2 + x - 12 = 0$

5) $x^2 + 7x - 18 = 0$

6) $x^2 - 2x - 15 = 0$

7) $x^2 + 6x - 40 = 0$

8) $x^2 - 9x - 36 = 0$

Graphing Quadratic Functions

Graphing a quadratic function involves plotting the points that satisfy the equation and then connecting them to form a parabola. The general form of a quadratic function is $y = ax^2 + bx + c$, where a, b, and c are constants.

The graph of a quadratic function is a parabola that opens either upward or downward, depending on the value of a. If a is positive, the parabola opens upward and if a is negative, the parabola opens downward. The vertex of the parabola is the point where it changes direction and it is given by the coordinates

$$\left(-\frac{b}{2a}, f\left(-\frac{b}{2a}\right)\right),$$

where $f(x)$ is the quadratic function.

The x-intercepts of the parabola are the solutions of the equation $y = 0$. We can find them by setting $y = 0$ and solving for x. They are given by

$$x = \frac{-b \pm \sqrt{b^2 - 4ac}}{2a}$$

To graph a quadratic function, we can use the vertex and the x-intercepts as key points and then plot several other points to complete the graph. Graphing calculators and software also can help to graph a quadratic function. (Remember that the graph of a quadratic function is a U-shaped curve and it is called a "parabola".)

Consider that quadratic functions in vertex form: $y = a(x - h)^2 + k$ where (h, k) is the vertex of the function. The axis of symmetry is $x = h$. The sign of a determines the direction of the corresponding graph. So that:

- If $a > 0$, then it opens upwards.
- If $a < 0$, then it opens downwards.

Let's review an example to better explanation:

Sketch the graph of $y = (x + 2)^2 - 3$.

The graph of $y = (x + 2)^2 - 3$ is a parabola that opens upward because a is positive. To sketch the graph, we can use the following steps:

Find the vertex: The vertex of the parabola is given by $(-2, -3)$.

Find the x−intercepts: We can set $y = 0$ and solve for x.

We will have $(x + 2)^2 = 3$, then $x^2 + 4x + 4 = 3$, so $x^2 + 4x + 1 = 0$, this equation has two negative solutions as

$x = \frac{-4 \pm \sqrt{4^2 - 4(1)(1)}}{2(1)} = \frac{-4 \pm \sqrt{12}}{2} = -2 \pm \sqrt{3}$.

Plot several points to complete the graph: we can pick several x−values, substitute them into the equation, and find the corresponding y−values.

For example, we can take $x = -5, -4, -3, -2, -1, 0$, and 1, and substitute them into the equation. We will have $y = 6, 1, -2, -3, -2, 1$, and 6 respectively.

Sketch the graph: Using the vertex and the points we found above, we can sketch the graph of the parabola. The vertex is the lowest point on the parabola and the parabola opens upward. The parabola doesn't cross the x−axis.

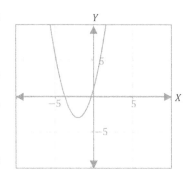

Note: The graph will be a parabola shape symmetric around $x = -2$.

✏ Draw the graph of the following functions.

9) $y = (x - 4)^2 - 2$

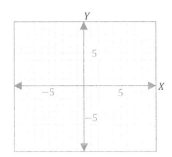

10) $y = 2(x + 2)^2 - 3$

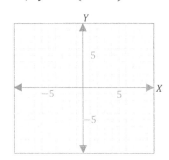

Solve a Quadratic Equation by Factoring

To solve a quadratic equation by factoring, we need to factor the left-hand side of the equation into the form of $(x - a)(x - b) = 0$, where a and b are the solutions of the equation. We can then set each factor equal to zero and solve for x.

For example, let's solve the equation $x^2 - 5x + 6 = 0$. We can factor it as:

$$x^2 - 5x + 6 = (x - 2)(x - 3) = 0$$

Now, we set each factor equal to zero and solve for x:

$$x - 2 = 0 \to x = 2 \text{ and } x - 3 = 0 \to x = 3.$$

So, the solutions of the equation $x^2 - 5x + 6 = 0 \to x = 2$ and $x = 3$.

Another example, let's solve the equation $x^2 + 2x - 8 = 0$. We can factor it as:

$$x^2 + 2x - 8 = (x + 4)(x - 2) = 0.$$

Now, we set each factor equal to zero and solve for x:

$$x + 4 = 0 \to x = -4 \text{ and } x - 2 = 0 \to x = 2.$$

So, the solutions of the equation $x^2 + 2x - 8 = 0$ are $x = -4$ and $x = 2$.

It's important to remember that factoring a quadratic equation requires finding two numbers that multiply to give the constant term and add to give the coefficient of x.

Here are some examples of solving a quadratic equation by factoring:

Solve the equation $x^2 - 8x + 12 = 0$.

Step 1: Factor the left-hand side of the equation:

$$x^2 - 8x + 12 = (x - 6)(x - 2) = 0.$$

Step 2: Set each factor equal to zero and solve for x:

$$x - 6 = 0 \to x = 6 \text{ and } x - 2 = 0 \to x = 2.$$

Step 3: Check the solution by substituting it back into the original equation:

$x = 6: (6)^2 - 8(6) + 12 = 36 - 48 + 12 = 0$

$x = 2: (2)^2 - 8(2) + 12 = 4 - 16 + 12 = 0$

So, the solutions of the equation $x^2 - 8x + 12 = 0$ are $x = 6$ and $x = 2$.

Find the solutions of $x^2 + 5x = 0$.

The greatest common factor of the two terms is x.

Take the common factor out: $x(x + 5) = 0$.

Using MPZ, which states that either $x = 0$ or $x + 5 = 0$.

In the second equation, the value of x equals -5. So, $x + 5 = 0 \to x = -5$.

To solve a quadratic equation as $ax^2 + bx + c = 0$ by factoring, if $a \neq 1$. In this case, it is enough to find two numbers whose product is ac and whose sum is b.

To better explain, pay attention to the following example:

Find the solutions of $2x^2 + 3x - 9 = 0$.

Since $a \neq 1$, determine two factors of the product of $ac = 2 \times (-9) = -18$ and add up to 3.

$2x^2 + 3x - 9 = 0 \to 2x^2 - 3x + 6x - 9 = 0 \to x(2x - 3) + 3(2x - 3) = 0$

$\to (2x - 3)(x + 3) = 0$

Lastly, by setting each factor to 0, we have:

$2x - 3 = 0 \to x = \frac{3}{2}$, or $x + 3 = 0 \to x = -3$.

So, the solutions of the equation $2x^2 + 3x - 9 = 0$ are $x = \frac{3}{2}$ and $x = -3$.

✏️ Find the solution of the following equations.

11) $x^2 - 2x - 3 = 0$

12) $x^2 + 9x + 20 = 0$

13) $6x^2 + x - 5 = 0$

14) $x^2 - 2x - 4 = 12$

15) $x^2 + 1 = 2x$

16) $x^2 - 16 = 0$

17) $5x^2 + 7x + 2 = 0$

18) $x^2 + x + 2 = 0$

Transformations of Quadratic Functions

Quadratic functions are functions of the form $f(x) = ax^2 + bx + c$, where a, b, and c are constants. The graph of a quadratic function is a parabola. The shape of the parabola can be changed by changing the values of a, b, and c. These changes are known as transformations of quadratic functions.

1. Vertical shifts: We can change the y-coordinate of the vertex of the parabola by adding or subtracting a constant to the function.

   $y = f(x) + k$: Moves up   if k is positive

   Moves down   if k is negative

2. Horizontal shifts: We can change the x-coordinate of the vertex of the parabola by adding or subtracting a constant to the x-term.

   $y = f(x + k)$: Moves left   if k is positive

   Moves right   if k is negative

3. Stretch or compression: We can change the width of the parabola by multiplying or dividing the x-term by a constant.

   $y = f(x) + k$: Compressed in direction y if $0 < k < 1$

   Stretched in direction y if $k > 1$

   As the same:

   $y = f(kx)$: Compressed in direction x if $0 < k < 1$

   Stretched in direction x if $k > 1$

4. Reflection: We can reflect the parabola over the x-axis or y-axis by changing the sign of the leading coefficient (a) or by multiplying the x-term by -1.

   $y = -f(x)$:   Reflection over x − axis

   $y = f(-x)$:   Reflection over y − axis

It's important to note that when we make these transformations, the vertex of the parabola will change. The vertex is the point at the bottom or top of the parabola. The standard form of the vertex form of a quadratic function is $y = a(x - h)^2 + k$, where (h, k) is the vertex of the parabola. By analyzing the vertex form we can understand the shift of the parabola.

Here are some examples of how to apply transformations to a quadratic function:

$$y = x^2 + 2x - 3$$

- To shift the parabola upward by 2 units, we add 2 to the y-coordinate:
$$y = x^2 + 2x - 3 + 2 \rightarrow y = x^2 + 2x - 1$$

- To shift the parabola to the left by 1 unit, we subtract 1 from the x-coordinate: $y = (x + 1)^2 + 2(x + 1) - 3 \rightarrow y = x^2 + 4x$.

- To stretch the parabola horizontally by a factor of 2, we multiply the x-term by 2: $y = (2x)^2 + 2(2x) - 3 \rightarrow y = 4x^2 + 4x - 3$.

- To reflect the parabola over the x-axis, we change the sign of the leading coefficient: $y = -(x^2 + 2x - 3) \rightarrow y = -x^2 - 2x + 3$.

🖎 State the transformations and sketch the graph of the following function.

19) $y = 2(x - 3)^2 + 1$

🖎 Write the equation for the function $y = x^2$ with the following transformations.

20) reflect across the x-axis and vertically compress by a factor of $\frac{3}{2}$

21) shift down 2

22) reflect across the y-axis, shift up 1

23) vertically stretch by a factor of 3, shift left 4 and up 1

Quadratic Formula and the Discriminant

The quadratic formula is a method for solving a quadratic equation of the form $ax^2 + bx + c = 0$. The formula is:

$$x = \frac{-b \pm \sqrt{b^2 - 4ac}}{2a}$$

The discriminant of the quadratic equation is the expression under the square root sign in the quadratic formula and is indicated by the notation Δ:

$$\Delta = b^2 - 4ac$$

The discriminant determines the nature of the solutions of the quadratic equation.

- If the discriminant is positive, the equation has two distinct real solutions.
- If the discriminant is zero, the equation has one repeated real solution.
- If the discriminant is negative, the equation has no real solutions.

In other words, the discriminant helps us understand the behavior of the graph of the equation, whether it has two distinct roots, one repeated root, or no real roots.

For more explanation, pay attention to the following examples:

Solve equation $4x^2 + 3x - 1 = 0$.

To solve a quadratic equation, first find the values of a, b, c.

By comparing the mentioned equation with the equation $ax^2 + bx + c = 0$, the values a, b, c are equal to: $a = 4, b = 3, c = -1$.

Now, calculate the discriminant of the equation (Δ). Given the values of a, b, and c, the value of Δ is equal to:

$\Delta = b^2 - 4ac = 3^2 - 4 \times 4 \times (-1) = 25$.

25 is a positive number. Therefore, this equation will have two

different solutions: $x = \frac{-3 \pm \sqrt{25}}{2(4)} \to x = \frac{-3 \pm 5}{8} \to x = -1$, or $x = \frac{1}{4}$.

Find the solutions of the equation $x^2 + x + 1 = 0$.

In the equation, the values of a, b, c are: $a = 1$, $b = 1$, $c = 1$.

$\Delta = b^2 - 4ac = 1^2 - 4 \times 1 \times 1 = 1 - 4 = -3$.

The value of the discriminant (Δ) is negative; Therefore, this equation has no solution in real numbers.

Solve the following quadratic equation:

$$x^2 + 4x + 4 = 0$$

In order to solve the problem, we evaluate the value of the discriminant (Δ). For this purpose, we determine the values of a, b, and c corresponding to the standard form of the quadratic equation $ax^2 + bx + c = 0$ and put them in the formula $\Delta = b^2 - 4ac$.

According to the given equation $x^2 + 4x + 4 = 0$, we get: $a = 1$, $b = 4$, and $c = 4$. Then, $\Delta = 4^2 - 4(1)(4) = 0$. So, the equation has one repeated real solution as:

$$x = \frac{-b}{2a} \to x = \frac{-4}{2(1)} \to x = -2$$

✎ How many solutions do the following equations have?

24) $x^2 + 2x + 4 = 0$

25) $x^2 - 4x - 5 = 0$

26) $3x^2 + x - 1 = 0$

27) $x^2 + 1 = 0$

✎ Find the answers to the questions below.

28) $2x^2 - 7x + 3 = 0$

29) $x^2 + 8x - 9 = 0$

30) $2x^2 + 5x - 3 = 0$

31) $x^2 + 6x + 9 = 0$

Characteristics of Quadratic Functions: Equations

By analyzing the equation of a quadratic function, we can determine various characteristics of the corresponding parabolic graph, including:

- Direction of the parabola: The direction of the parabola refers to whether the parabola opens upwards or downwards. This can be determined by the sign of the leading coefficient 'a' in the equation of the quadratic function. If 'a' is positive, the parabola opens upwards, and if 'a' is negative, the parabola opens downwards.
- Vertex: The vertex of a parabola is the highest or lowest point of the graph, and it can be determined by using the vertex form of the equation:

$$y = a(x - h)^2 + k,$$

  where (h, k) is the vertex of the parabola or using formula $\left(-\frac{b}{2a}, f\left(-\frac{b}{2a}\right)\right)$ for the standard form of a quadratic equation $y = ax^2 + bx + c$.
- Axis of symmetry: The axis of symmetry of a parabola is a vertical line that separates the parabola into two mirror-image halves. It can be determined by using the vertex form of the equation, the axis of symmetry is the line $x = h$, where h is the $x-$coordinate of the vertex.
- $x-$intercept(s): The $x-$intercepts of a quadratic function are the points at which the parabola crosses the $x-$axis. These can be found by setting $y = 0$ and solving for x.
- $y-$intercept: The $y-$intercept of a quadratic function is the point at which the parabola crosses the $y-$axis, this point is represented by $(0, c)$ in the equation $y = ax^2 + bx + c$.
- Range: The range of a quadratic function is the set of all possible $y-$values of the function. It can be determined by analyzing the vertex and the direction of the parabola.
  - Minimum/maximum value: The minimum or maximum value of a

quadratic function can be found by using the vertex form of the equation, it is the y-coordinate of the vertex (h, k), and it indicates the lowest or highest value of the function, depending on the direction of the parabola. If the parabola opens upwards, the vertex represents the minimum value, and if it opens downwards, the vertex represents the maximum value. This can also be determined by looking at the leading coefficient, if a is positive, the vertex is a minimum point, if a is negative, the vertex is a maximum point.

Let's review an example:

Specify characteristics of the vertex, direction, and y-intercept for the quadratic function $f(x) = x^2 + 3x + 2$.

In the standard form of a quadratic function $f(x) = ax^2 + bx + c$, the vertex is not immediately obvious. The x-coordinate for the vertex can be obtained by using the formula $x = -\frac{b}{2a}$. So, $x = -\frac{3}{2}$.

Now, substitute it into the equation of function to obtain the y-coordinate:

$$y = f\left(-\frac{3}{2}\right) = \left(-\frac{3}{2}\right)^2 + 3\left(-\frac{3}{2}\right) + 2 = \left(\frac{9}{4}\right) - \frac{9}{2} + 2 = -\frac{1}{4}.$$

So, the vertex of the function $f(x) = x^2 + 3x + 2$ is the ordered pair $\left(-\frac{3}{2}, -\frac{1}{4}\right)$.

Remember that the sign of the coefficient x^2 indicates the direction of the quadratic equation. Since the coefficient of x^2 is $+1$, then it is upward.

To find the y-intercept of a function, evaluate the output at $f(0)$.

$$f(0) = (0)^2 + 3(0) + 2 = 2.$$

✏ Solve.

32) Find the equation of the axis of symmetry for the parabola $y = x^2 + 7x + 3$.

33) Find the y-intercept of the parabola $y = x^2 + 25x + 7$.

34) Find the vertex of the parabola $y = x^2 - 4x + 3$.

Characteristics of Quadratic Functions: Graphs

When graphed, quadratic functions take the shape of a parabola, which is a symmetric, U-shaped curve. The graph of a quadratic function has several key characteristics that can be determined from the equation of the function:

- Vertex: The vertex is the highest or lowest point of the parabola.
- Axis of Symmetry: The axis of symmetry of a parabola is a vertical line that separates the parabola into two mirror-image halves. The axis of symmetry is the line $x = h$, where h is the x-coordinate of the vertex.
- x-intercepts: The x-intercepts are the points at which the parabola crosses the x-axis.
- y-intercept: The y-intercept is the point at which the parabola crosses the y-axis, it is represented by $(0, c)$ in the equation $y = ax^2 + bx + c$.
- Minimum/maximum: The minimum or maximum value of a quadratic function is equivalent to the vertex of the function, which can be clearly identified from the graph.
- Increase/decrease: According to the parabolic form of a quadratic function, the increasing or decreasing of the graph at the vertex changes according to the sign of the leading coefficient of the equation. This property can be seen in the graph.
- Concavity: The parabola of a quadratic function is either concave up or concave down, depending on the sign of the leading coefficient of the equation. If the leading coefficient is positive, the parabola is concave up, and if it is negative, the parabola is concave down. The direction of concavity of a parabola can be determined by evaluating the graph.

All of these characteristics can be used to analyze and interpret the behavior of a quadratic function and its corresponding graph. The shape and position of the graph can give insight into the properties of the function and how it behaves in different situations. Look at the following graphs to understand more:

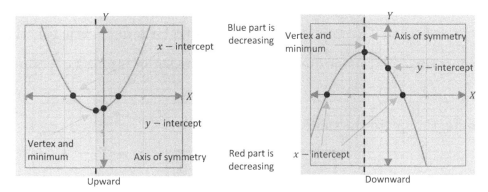

For example:

Considering the following graph, determine the following characteristics: vertex, axis of symmetry, x-intercepts, y-intercept, Max/minimum point, concavity

According to the graph, see that a point at the coordinate $(2,4)$ is the vertex. So, the line $x = 2$ is the axis of symmetry.

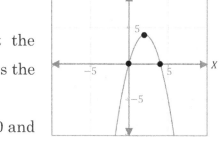

Since the graph intersects the x-axis at points 0 and 4, the mentioned points are x-intercepts. In the same way, the point $(0,0)$ is the y-intercept.

In addition, this is clear that the vertex is the maximum point. Clearly, the graph is downward.

✎ Considering the following graph, determine the following:

35) vertex

36) axis of symmetry

37) y-intercepts

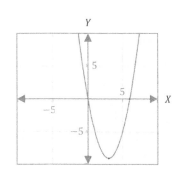

Complete a Function Table: Quadratic Functions

A function table is a tool that can be used to organize and display the inputs and outputs of a function. To complete a function table for a quadratic function, you need to know the equation of the function and the set of inputs (x –values) that you want to use.

For example: Complete the table.

$g(x) = -2x^2 + 4x + 1$	
x	$f(x)$
-2	
0	
1	
4	

According to the function table, the first value of the input in the table is -1. Evaluate $g(x) = -2x^2 + 4x + 1$ for $t = -2$.

$g(-2) = -2(-2)^2 + 4(-2) + 1$
$\quad\quad\quad = -2(4) - 8 + 1 = -15$

When $t = -2$, then $g(-2) = -15$.

$g(x) = -2x^2 + 4x + 1$	
x	$f(x)$
-2	-15
0	
1	
4	

Complete the first row of the table.

In the same way, evaluate $g(x) = -2x^2 + 4x + 1$ for $t = 0$, $t = 1$, and $t = 4$, respectively. So,

$g(0) = -2(0)^2 + 4(0) + 1 = 1$
$g(1) = -2(1)^2 + 4(1) + 1 = 3$
$g(4) = -2(4)^2 + 4(4) + 1 = -15$

Enter the obtained values in the table.

$g(x) = -2x^2 + 4x + 1$	
x	$f(x)$
-2	-15
0	1
1	3
4	-15

bit.ly/3XlOVuF
Find more at

Here's another example:

Let's say we have the quadratic function: $y = x^2 + 2x - 3$.

And we want to create a function table with x-values of $-2, -1, 0,$ and 3. The completed function table would look like this:

$f(x) = x^2 + 2x - 3$	
x	$f(x)$
-2	-3
-1	-4
0	-3
3	12

To complete the table, we substitute each x-value into the equation and solve for y. You can use a similar approach for any quadratic function, just substitute the equation you have into the table, and for each x-value in the table, substitute it in the equation and solve for y.

✎ Complete the following tables.

38)

$g(t) = t^2 + 7$	
t	$g(t)$
-1	
0	
1	

39)

$f(p) = 4p^2$	
p	$f(p)$
-2	
0	
2	

Domain and Range of Quadratic Functions: Equations

The domain of a function is the set of all possible input values (x –values) for which the function produces a valid output (y –value). The range of a function is the set of all possible output values (y –values) that the function can produce.

For a quadratic function in the form of $y = ax^2 + bx + c$, the domain is all real numbers, because any value of x can be plugged into the equation and produce a valid output.

The range of a quadratic function can be found by examining the leading coefficient of the equation.

- If the leading coefficient (a) is positive, the parabola opens upwards and the vertex is a minimum point, the range is all real numbers greater than the y –coordinate of the vertex.
- If the leading coefficient (a) is negative, the parabola opens downwards and the vertex is a maximum point, the range is all real numbers less than the y –coordinate of the vertex.

For example: What is the domain and range of the function to the equation?

$$y = x^2 + 2x + 1$$

The function is quadratic, then the domain is all real numbers. Because every value of x is a valid input of the quadratic function. Now, to find the range of the function, rewrite the equation as a perfect square. That is, $y = (x + 1)^2$.

Since the value of the expression $(x - 1)^2$ is always greater than and equal to 0. Then the range of the function includes all non-negative real numbers.

Here's another example:

If we have the quadratic function $y = 2x^2 - 4x - 6$, the leading coefficient is positive, so the parabola opens upward and the vertex is a minimum point. By completing the square, we can find that the vertex is located at $(1, -8)$, which is the minimum point of the graph. Therefore, the range is all real numbers greater than -8.

bit.ly/3ZQQsuH
Find more at

More examples:

Let's say we have the quadratic function: $y = -2x^2 - 4x + 5$.

In this equation, the leading coefficient (a) is negative, so the parabola opens downward and the vertex is a maximum point. By completing the square, we can find that the vertex is located at $(-1, 7)$, which is the maximum point of the graph. Therefore, the range is all real numbers less than 7.

So, the domain of this function is all real numbers, and the range is all real numbers less than 7.

It's good to note that in some cases, the range may not be restricted to a certain set of numbers. In those cases, the range would be all real numbers.

Find the range of the function $y = x^2 - 5x + 3$.

Rewrite the equation of the function as a perfect square. So,

$$x^2 - 5x + 3 = \left(x^2 - 5x + \frac{25}{4} - \frac{25}{4}\right) + 3 = \left(x - \frac{5}{2}\right)^2 - \frac{25}{4} + 3 = \left(x - \frac{5}{2}\right)^2 - \frac{13}{4}.$$

According to the obtained equation, since the domain of the function is all real numbers, so, we have: $\left(x - \frac{5}{2}\right)^2 \geq 0 \rightarrow \left(x - \frac{5}{2}\right)^2 - \frac{13}{4} \geq -\frac{13}{4}$. Therefore, the range of the function is the set of $\left\{y \in \mathbb{R} | y \geq -\frac{13}{4}\right\}$.

Evaluate the range of the following function: $y = -3x^2 + x - 1$.

Rewrite the equation of function as

$$-3x^2 + x - 1 = -3\left(x^2 - \frac{1}{3}x\right) - 1 = -3\left(x^2 - \frac{1}{3}x + \frac{1}{36} - \frac{1}{36}\right) - 1 = -3\left(x - \frac{1}{3}\right)^2 - 3\left(-\frac{1}{36}\right) - 1 = -3\left(x - \frac{1}{3}\right)^2 - \frac{11}{12}.$$

Since $\left(x - \frac{1}{3}\right)^2 \geq 0 \rightarrow -3\left(x - \frac{1}{3}\right)^2 \leq 0 \rightarrow -3\left(x - \frac{1}{3}\right)^2 - \frac{11}{12} \leq -\frac{11}{12} \rightarrow y \leq -\frac{11}{12}.$

Therefore, the range of the function is the set of $\left\{y \in \mathbb{R} | y \leq -\frac{11}{12}\right\}$.

✎ Find the domain and range.

40) $y = x^2 + 5x + 6$

41) $y = -x^2 + 4$

Factor Quadratics: Special Cases

The special cases for factoring quadratics are when the equation has a square or difference of squares.

- When factoring a square, the equation is of the form $a^2 \pm 2ab + b^2$. In this case, you can factor it as $(a \pm b)^2$.
- When factoring a difference of squares, the equation is of the form $a^2 - b^2$. In this case, you can factor it as $(a + b)(a - b)$.

It's important to note that these special cases only work when the equation is in a specific form and you cannot always factor a quadratic equation using this method.

Here are a couple of examples:

Factor the equation $9x^2 - 36$.

Notice that $9x^2$ and 36 are perfect squares, because $9x^2 = (3x)^2$, and $36 = 6^2$. By using the formula $a^2 - b^2 = (a + b)(a - b)$.

Let $a = 3x$, and $b = 6$. The represented equation can be rewritten as follow:

$9x^2 - 36 = (3x + 6)(3x - 6) = 9(x + 2)(x - 2)$.

Factor $25n^2 - 10n + 1$.

First, consider that this formula $a^2 - 2ab + b^2 = (a - b)^2$. Since $25n^2 = (5n)^2$, then $25n^2$ and 1 are perfect squares. This means that $a = 5n$, and $b = 1$. Next, check to see if the middle term is equal to $2ab$, which it is: $2ab = 2(5n)(1) = 10n$. Therefore, the square equation can be rewritten as,

$25n^2 - 10n + 1 = (5n - 1)^2$.

Factor $49\alpha^2 + 28\alpha\beta + 4\beta^2$.

First, notice that $49\alpha^2$ and $4\beta^2$ are perfect squares because $49\alpha^2 = (7\alpha)^2$ and $4\beta^2 = (2\beta)^2$. Let $a = 7\alpha$, and $b = 2\beta$. Now, evaluate the middle term

$28\alpha\beta$. Then, you can write as $28\alpha\beta = 2(7\alpha)(2\beta) = 2ab$. Using this formula:
$$a^2 + 2ab + b^2 = (a+b)^2.$$

Therefore, we have: $49\alpha^2 + 28\alpha\beta + 4\beta^2 = (7\alpha + 2\beta)^2$.

Factor the equation $3x^2 - 12x + 12$.

First, rewrite the equation as $3x^2 - 12x + 12 = 3(x^2 - 4x + 4)$. According to the x^2 and 4 are perfect squares. That is, $a = x$, and $b = 2$. Now, check to see if the middle term is equal to $2ab$: $2ab = 2(x)2 = 4x$. Using this formula: $a^2 - 2ab + b^2 = (a-b)^2$.

Therefore, we have: $3x^2 - 12x + 12 = 3(x^2 - 4x + 4) = 3(x-2)^2$.

Factor the equation $5x^2 - 80$.

By factoring from 5, rewrite the equation as $5x^2 - 90 = 5(x^2 - 16)$. We see that this expression in the parentheses is a difference of squares because it can be written as $(x+4)(x-4)$, so the factored form of this equation is: $5x^2 - 80 = 5(x+4)(x-4)$.

Expand the expression: $(x+5)^2$.

Using this formula: $(a+b)^2 = a^2 + 2ab + b^2$. So, the expnded form of this equation is $(x+5)^2 = x^2 + 10x + 25$.

✎ Factor the following expressions.

42) $25x^2 + 20x + 4$

43) $9x^2 - 1$

44) $3 + 6x + 3x^2$

45) $b^4 - 36$

✎ Expand the following expressions.

46) $(2-x)(x+2)$

47) $(x-2y)(x+2y)$

48) $(7-2r)^2$

49) $(x+2)(x+2)$

Factor Quadratics Using Algebra Tiles

To factor quadratic expressions like $ax^2 + bx + c$ using algebraic tiles, follow the steps below:

- Model the polynomials with tiles.
- Arrange the tiles into a rectangle grid. Start with the x^2 tiles from the upper left corner so that the number of horizontal and vertical divisions is equal to the multiples of a. Add the integer tiles in the lower right corner. Here, the number of horizontal and vertical divisions should be equal to integer multiples. Make sure to choose horizontal and vertical divisions from possible multiples of a and c that will fill the remaining empty grid tiles.
- The product of expressions related to horizontal and vertical divisions is equal to the answer.
- Check the answer.

For example, to better explanation:

Use algebra tiles to factor: $3x^2 - 7x + 2$.

Model the polynomials with tiles:

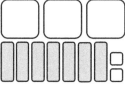

In this case, arrange the tiles into a rectangle grid.

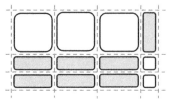

Determine both binomials relate to the divisions, such that $3x - 1$ for the horizontal division and $x - 2$ for the vertical. As follow:

In the end, multiply two expressions and check the answer:

$(3x - 1)(x - 2) = 3x^2 - 7x + 2$.

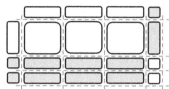

Use algebra tiles to factor $2x^2 + x - 3$.

Represent this expression using the number of 2 tiles x^2 ($2x^2$), and 1 tile x, and 3 negative integer tiles (-3).

In this case, arrange the tiles into a rectangle grid. Note that there aren't enough x tiles to correctly fit all of the integer tiles into the larger rectangle.

To fix this problem, we need to add some tiles in the empty spaces of the grid so that the sum of the positive and negative tiles is zero. Adding 2 positive x tiles and 2 negative x tiles does not change the value of the original expression. Since $-2x + 2x = 0$. These pairs cancel the combination. Make sure you place the negative x tiles on one side of the x^2 tile and all the positive x tiles on the other side.

Determine both binomials relate to the divisions, such that $2x + 3$ for the horizontal division and $x - 1$ for the vertical. Finally, check the answer.

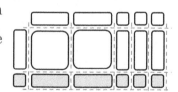

$(2x + 3)(x - 1) = 2x^2 + x - 3$.

✎ Use algebra tiles to factor the following equations.

50) $x^2 - 3x + 2$ 51) $x^2 + 5x + 6$

Write a Quadratic Function from Its Vertex and Another Point

To write a quadratic function from its vertex and another point, you can use the vertex form of the equation which is $y = a(x - h)^2 + k$.

Here, (h, k) is the vertex of the parabola and a is the leading coefficient of the equation.

You can use the following steps to write the quadratic function:

1. Identify the vertex of the parabola: The vertex is the turning point of the parabola, and it is the point of minimum or maximum value.
2. Identify the other point: This can be any point on the parabola.
3. Substitute the coordinates of the vertex and the other point into the vertex form of the equation: $y = a(x - h)^2 + k$.
4. Solve for a: Once you have the equation set up with the vertex and another point, you can solve for the leading coefficient.
5. Write the final equation: Once you have solved for a, you can write the final equation in the form $y = a(x - h)^2 + k$.

For example, if the vertex of the parabola is $(-1, 2)$ and the other point on the parabola is $(1, 6)$, the steps would be:

1. Identify the vertex: $(-1, 2)$.
2. Identify the other point: $(1, 6)$.
3. Substitute the coordinates into the vertex form: $y = a(x + 1)^2 + 2$.
4. Solve for a: $6 = a(1 + 1)^2 + 2 \rightarrow 6 = 4a + 2 \rightarrow a = 1$.
5. Write the final equation: $y = 1(x + 1)^2 + 2$.

So, the final equation is $y = (x + 1)^2 + 2$.

Let's review an example:

A quadratic function opening up or down has a vertex $(0,-4)$ and passes through $(4,0)$. Write its equation in vertex form.

Use the vertex form of the quadratic function as $y = a(x-h)^2 + k$. Substitute the coordinate of the vertex $(0,-4)$ in the vertex form:

$y = a(x-0)^2 + (-4) \to y = ax^2 - 4$.

To find a substitute $(4,0)$ in this equation and calculate. Then,

$(4,0) \to 0 = a(4-0)^2 - 4 \to 0 = 16a - 4 \to 16a = 4 \to a = \frac{1}{4}$.

Therefore, the equation of the quadratic function in the vertex form is as follows:

$y = \frac{1}{4}x^2 - 4$.

A quadratic function has a vertex $\left(-\frac{1}{4}, -\frac{25}{8}\right)$ and passes through $(2,7)$. Write its equation in vertex form.

By using the vertex form formula: $y = a(x-h)^2 + k$. So, we have:

$\left(-\frac{1}{4}, -\frac{25}{8}\right) \to y = a\left(x - \left(-\frac{1}{4}\right)\right)^2 + \left(-\frac{25}{8}\right) \to y = a\left(x + \frac{1}{4}\right)^2 - \frac{25}{8}$.

Substitute $(2,7)$ in the obtained equation, then:

$7 = a\left(2 + \frac{1}{4}\right)^2 - \frac{25}{8} \to 7 = \frac{81}{16}a - \frac{25}{8} \to \frac{81}{16}a = \frac{81}{8} \to a = 2$.

Therefore, $y = 2\left(x + \frac{1}{4}\right)^2 - \frac{25}{8}$.

✎ Find the quadratic equations that passes through the given vertex and another point.

52) A parabola opening or down has vertex $(0,0)$ and passes through $(8,-16)$.

53) A parabola opening up or down has a vertex $(0,2)$ and passes through $(-2,5)$.

Chapter 6: Answers

1) $x = 2, x = -1$
2) $x = 2, x = 4$
3) $x = 3, x = 1$
4) $x = 3, x = -4$
5) $x = 2, x = -9$
6) $x = 5, x = -3$
7) $x = 4, x = -10$
8) $x = 12, x = -3$

9)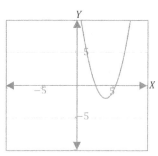

10)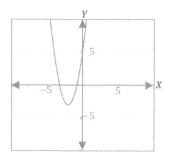

11) $\{-1, 3\}$
12) $\{-5, -4\}$
13) $\{-1, \frac{5}{6}\}$
14) $\{1 - \sqrt{17}, 1 + \sqrt{17}\}$
15) $\{1\}$
16) $\{-4, 4\}$
17) $\{-1, -\frac{2}{5}\}$
18) No real solution

19) The graph stretches vertically by a factor of 2. Move 3 units to the right and 1 unit up.

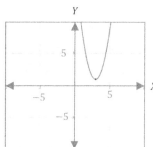

20) $y = -\frac{2}{3}x^2$
21) $y = x^2 - 2$
22) $y = x^2 + 1$
23) $y = 3(x + 4)^2 + 1$
24) No solution
25) Two different solutions
26) Two different solutions

27) No solution
28) $x_1 = 3, x_2 = \frac{1}{2}$
29) $x_1 = -9, x_2 = 1$
30) $x_1 = -3, x_2 = \frac{1}{2}$
31) $x_1 = x_2 = -3$

32) $x = -\frac{7}{2}$
33) 7
34) $(2, -1)$
35) $(3, -9)$
36) $x = 3$
37) 0

38)

$g(t) = t^2 + 7$	
t	$g(t)$
-1	8
0	7
1	8

39)

$f(p) = 4p^2$	
p	$f(p)$
-2	16
0	0
2	16

40) $D = \{x | x \in R\}$,
$R = \{y \in R | y \geq -\frac{1}{4}\}$

41) $D = \{x | x \in R\}$,
$R = \{R | y \leq 4\}$

42) $(5x + 2)^2$
43) $(3x - 1)(3x + 1)$
44) $3(x + 1)^2$
45) $(b^2 + 6)(b^2 - 6)$
50) $x^2 - 3x + 2 = (x - 1)(x - 2)$

46) $-x^2 + 4$
47) $x^2 - 4y^2$
48) $4r^2 - 28r + 49$
49) $x^2 + 4x + 4$
51) $x^2 + 5x + 6 = (x + 2)(x + 3)$

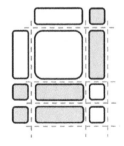
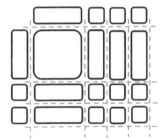

52) $y = -\frac{1}{4}x^2$
53) $y = \frac{3}{4}x^2 + 2$

Chapter 7: Polynomials

Math topics that you'll learn in this chapter:

- ☑ Simplifying Polynomials
- ☑ Adding and Subtracting Polynomials
- ☑ Add and Subtract Polynomials Using Algebra Tiles
- ☑ Multiplying Monomials
- ☑ Dividing Monomials
- ☑ Multiplying a Polynomial and a Monomial
- ☑ Multiply Polynomials Using Area Models
- ☑ Multiplying Binomials
- ☑ Multiply two Binomials Using Algebra Tiles
- ☑ Factoring Trinomials
- ☑ Factoring Polynomials
- ☑ Use a Graph to Factor Polynomials
- ☑ Factoring Special Case Polynomials
- ☑ Add Polynomials to Find Perimeter

Simplifying Polynomials

Simplifying polynomials is the process of combining like terms and reducing the degree of the polynomial.

To simplify a polynomial, you can follow these steps:

Step 1: Identify the like terms: These are terms that have the same variables raised to the same powers. For example, $3x^2$ and $5x^2$ are like terms.

Step 2: Combine the like terms: Add or subtract the numerical coefficients of the like terms. For example, you can combine $3x^2$ and $5x^2$ to get $8x^2$.

Step 3: Write the polynomial in descending order of exponents: This is done by arranging the terms in decreasing order of exponents, starting with the term with the highest exponent.

Step 4: Check for any remaining like terms, and combine them.

Step 5: Simplify any constants or coefficients that can be simplified.

For example, to simplify the polynomial $3x^2 + 2x^2 + 5x - 7x - 4x + 6$,

Step 1: Identify the like terms: $3x^2$ and $2x^2$, $5x$, $-7x$, and $-4x$.

Step 2: Combine the like terms: $3x^2 + 2x^2 = 5x^2$, $5x - 7x - 4x = -6x$.

Step 3: Write the polynomial in descending order of exponents: $5x^2 - 6x + 6$.

Step 4: Check for any remaining like terms and combine them: $5x^2 - 6x + 6$ (No more like terms).

Step 5: Simplify any constants or coefficients that can be simplified: $5x^2 - 6x + 6$ (No constants or coefficients can be simplified).

So, the simplified polynomial is $5x^2 - 6x + 6$.

Here's another example:

Simplify the polynomial $2x^3 - x + 5x^2 - 3x^2 - 8 + 6x$.

Step 1: Identify the like terms: $5x^2$ and $-3x^2$, $6x$ and $-x$.

Step 2: Combine the like terms: $5x^2 - 3x^2 = 2x^2$, and $6x - x = 5x$.

Step 3: Write the polynomial in descending order of exponents:

Polynomials

$$2x^3 + 2x^2 + 5x - 8.$$

Step 4: Check for any remaining like terms and combine them:

$$2x^3 + 2x^2 + 5x - 8 \text{ (No more like terms)}.$$

Step 5: Simplify any constants or coefficients that can be simplified:

$$2x^3 + 2x^2 + 5x - 8$$

(No constants or coefficients can be simplified).

So, the simplified polynomial is $2x^3 + 2x^2 + 5x - 8$.

Some more examples:

Simplify this expression. $3x(2x - 5) + 7x =$

Use Distributive Property: $3x(2x - 5) = 6x^2 - 15x$.

Now, combine like terms:

$$3x(2x - 5) + 7x = 6x^2 - 15x + 7x = 6x^2 - 8x.$$

Simplify this expression. $(x + 3)(x - 5) - 2x^2 + 2x =$.

First, apply the FOIL method:

$$(a + b)(c + d) = ac + ad + bc + bd.$$

Therefore, use the Distributive Property: $(x + 3)(x - 5) = x^2 - 5x + 3x - 15$.

Next, substitute the obtained expression in the original expression:

$$(x + 3)(x - 5) - 2x^2 + 2x = x^2 - 5x + 3x - 15 - 2x^2 + 2x.$$

Now combine like terms: $x^2 - 2x^2 = -x^2$, and $-5x + 3x + 2x = 0$.

The simplified form of the expression: $(x + 3)(x - 5) - 2x^2 + 2x = -x^2 - 15$.

✏️ Simplify the following expressions.

1) $3(6x + 4) =$

2) $5(3x - 8) =$

3) $x(7x + 2) + 9x =$

4) $6x(x + 3) + 5x =$

5) $6x(3x + 1) - 5x =$

6) $x(3x - 4) + 3x^2 - 6 =$

7) $x^2 - 5 - 3x(x + 8) =$

8) $2x^2 + 7 - 6x(2x + 5) =$

Adding and Subtracting Polynomials

Adding and subtracting polynomials involves combining like terms, just as simplifying polynomials does. The process is the same, but instead of simplifying a single polynomial, you are combining two or more polynomials.

For example, to add the polynomials $2x^3 + 3x^2 - x - 8$ and $3x^3 - 4x^2 + 8x + 4$:

1. Identify the like terms: $2x^3$ and $3x^3$, $3x^2$ and $-4x^2$, $-x$ and $8x$, -8 and 4.
2. Add the like terms together:
   $$2x^3 + 3x^3 = 5x^3, 3x^2 - 4x^2 = -x^2, -x + 8x = 7x, \text{ and } -8 + 4 = -4.$$
3. Write the polynomial in descending order of exponents: $5x^3 - x^2 + 7x - 4$.
4. Check for any remaining like terms and combine them: $5x^3 - x^2 + 7x - 4$ (No more like terms).
5. Simplify any constants or coefficients that can be simplified:
   $$5x^3 - x^2 + 7x - 4$$
   (No constants or coefficients can be simplified).

So, the sum of the polynomials is $5x^3 - x^2 + 7x - 4$.

You can do the same process of subtracting polynomials, by replacing the addition with subtraction of the like terms.

Let's another example:
$$(4x^4 - 3x^3 + 2x^2 - x - 6) - (2x^4 + x^3 - 5x^2 + 4x + 2)$$

To subtract the polynomials $4x^4 - 3x^3 + 2x^2 - x - 6$ and $2x^4 + x^3 - 5x^2 + 4x + 2$:

1. Identify the like terms:
   $4x^4$ and $2x^4$, $-3x^3$ and x^3, $2x^2$ and $-5x^2$, $-x$ and $4x$, -6 and 2.
2. Subtract the like terms: $4x^4 - 2x^4 = 2x^4$, $-3x^3 - x^3 = -4x^3$, $2x^2 - (-5x^2) = 7x^2$, $-x - 4x = -5x$, $-6 - 2 = -8$.
3. Write the polynomial in descending order of exponents:
   $2x^4 - 4x^3 + 7x^2 - 5x - 8$.

4. Check for any remaining like terms and combine them:

$2x^4 - 4x^3 + 7x^2 - 5x - 8$ (No more like terms).

5. Simplify any constants or coefficients that can be simplified:

$2x^4 - 4x^3 + 7x^2 - 5x - 8$

(No constants or coefficients can be simplified).

So, the subtraction of the polynomials is $2x^4 - 4x^3 + 7x^2 - 5x - 8$.

Simplify the expressions. $(5 + x^2 - 2x^3) - (x^3 - 3x^2 - 3) =$

First, use the Distributive Property:

$-(x^3 - 3x^2 - 3) = -x^3 + 3x^2 + 3$.

Substitute the last expression in the given expression:

$(5 + x^2 - 2x^3) - (x^3 - 3x^2 - 3) = 5 + x^2 - 2x^3 - x^3 + 3x^2 + 3$.

Now, combine like terms: $-2x^3 - x^3 = -3x^3$, $x^2 + 3x^2 = 4x^2$ and $5 + 3 = 8$.

Then: $(5 + x^2 - 2x^3) - (x^3 - 3x^2 - 3) = -3x^3 + 4x^2 + 8$.

Simplify the expressions. $(8x^2 - 3x^3) + (2x^2 + 5x^3) =$

Since the operation between the parentheses is summation, remove parentheses:

$(8x^2 - 3x^3) + (2x^2 + 5x^3) = 8x^2 - 3x^3 + 2x^2 + 5x^3$.

Now, combine like terms and write in standard form:

$8x^2 - 3x^3 + 2x^2 + 5x^3 = 10x^2 + 2x^3 = 2x^3 + 10x^2$.

✒ Simplify the following expressions.

9) $(x^2 + 3) + (2x^2 - 4) =$

10) $(3x^2 - 6x) - (x^2 + 8x) =$

11) $(4x^3 - 3x^2) + (2x^3 - 5x^2) =$

12) $(6x^3 - 7x) - (5x^3 - 3x) =$

13) $(10x^3 + 4x^2) + (14x^2 - 8) =$

14) $(4x^3 - 9) - (3x^3 - 7x^2) =$

15) $(9x^3 + 3x) - (6x^3 - 4x) =$

16) $(7x^3 - 5x) - (3x^3 + 5x) =$

Add and Subtract Polynomials Using Algebra Tiles

To better understand and visualize the addition and subtraction of algebraic expressions, you can use Algebra tiles as follows:

- Model the polynomials using tiles.
- For algebraic subtraction, change the color of the tiles on the second side.
- Cross out the same number of negative or positive tiles on both sides of the equation.
- Write the answer by determining the number of remaining tiles.

Let's review an example:

Use algebra tiles to simplify: $(x^2 - x + 3) + (2x^2 + 3x - 2)$.

Model the given polynomials using algebra tiles.

Here, cross out one x tile on the left side and do the same on the other side. In the same way, cancel two 1 tiles on the left side and do the same on the right side.

That is, count the number of remaining tiles.

So, $3x^2 + 2x + 1$.

Simplify the polynomial $(2x^2 + 3x - 1) - (x^2 - 2x - 2)$ using algebra tiles.

Model the polynomials with tiles.

Change the color of the tiles on the second side, then add them to the first side.

Polynomials

Now, simplify the obtained algebraic tiles by canceling negative and positive tiles of the same size. As follow:

Finally, by counting the remaining tiles, the following expression obtains:

$$x^2 + 5x + 1$$

One more example:

Use algebra tiles to simplify: $(2x^2 - 5x + 3) - (x^2 - 3x + 6)$.

Model the given expression using algebra tiles.

Change the color of the tiles related to the second side to show the negative sign applied to the second side term, then add them to the first side:

Next, cancel the same tiles with the different colors (negative and positive tiles) from the sides.

Now, count the number of remaining tiles.

$$x^2 - 2x - 3$$

✎ Use algebra tiles to simplify the following expressions.

17) $(2x^2 - 3x + 3) - (x^2 - x - 1)$

18) $(2x^2 + 2x + 5) + (x^2 + 2x + 1)$

Multiplying Monomials

When multiplying monomials, the process involves first multiplying the coefficients and then using the multiplication property of exponents for the variables.

- A monomial is a polynomial with just one term. For example, $2x$ or $7x^2$.
- When you multiply monomials, first multiply the coefficients (a number placed before and multiplying the variable) and then multiply the variables using the multiplication property of exponents.

$$x^a \times x^b = x^{a+b}$$

To illustrate this, consider the example of multiplying $3x^2$ and $4x^3$, you would first multiply the coefficients $3 \times 4 = 12$, then add the exponents of x, $2 + 3 = 5$, resulting in the product $12x^5$ ($x^2 \times x^3 = x^{2+3} = x^5$).

Here are a few examples with explanations:

$$(2x)(3x) = 6x^2$$

In this case, we have two monomials, $2x$ and $3x$. To multiply them, we simply multiply the numerical coefficients (2 and 3), which is $2 \times 3 = 6$.

Next, combine the variables (x) by using the formula $x^a \times x^b = x^{a+b}$ as

$$x \times x = x^{1+1} = x^2$$

The result is $6x^2$, which is a monomial with a numerical coefficient of 6 and a variable of x raised to the power of 2.

$$(-5y)(2y) = -10y^2$$

In this case, we have two monomials, $-5y$ and $2y$. To multiply them, we again multiply the numerical coefficients (-5 and 2) and combine the variables (y). The result is $-10y^2$, which is a monomial with a numerical coefficient of -10 and a variable of y raised to the power of 2.

$$(4)(-3) = -12$$

In this case, we have two monomials, 4 and -3. To multiply them, we again multiply the numerical coefficients (4 and -3) and combine the variables. The result is -12, which is a monomial with a numerical coefficient of -12 and no variable.

$$(2x^2)(3x^3) = 6x^5$$

In this case, we have two monomials, $2x^2$ and $3x^3$. To multiply them, we again multiply the numerical coefficients (2 and 3) and combine the variables (x) with their exponents (2 and 3) as follows: $x^2 \times x^3 = x^{2+3} = x^5$.

The result is $6x^5$, which is a monomial with a numerical coefficient of 6 and a variable of x raised to the power of 5.

Some more examples:

Multiply expressions. $2xy^3 \times 6x^4y^2$.

Find the same variables (x and x^4), and (y^3 and y^2). Then, use the multiplication property of exponents for multiplying the same variables: $x^a \times x^b = x^{a+b}$.

So, we have: $x \times x^4 = x^{1+4} = x^5$ and $y^3 \times y^2 = y^{3+2} = y^5$.

Then, multiply coefficients and variables: $2xy^3 \times 6x^4y^2 = 12x^5y^5$.

Multiply. $-5x^2z^3 \times 4z^5y^7$.

Use the multiplication property of exponents: $x^a \times x^b = x^{a+b}$. First multiply the same variable(s): $z^3 \times z^5 = z^{3+5} = z^8$. Then: $-5x^2z^3 \times 4z^5y^7 = -20x^2y^7z^8$.

Simplify. $(-6a^7b^4)(4a^8b^5) =$

Use the multiplication property of exponents: $x^a \times x^b = x^{a+b}$. $a^7 \times a^8 = a^{7+8} = a^{15}$ and $b^4 \times b^5 = b^{4+5} = b^9$. Then: $(-6a^7b^4)(4a^8b^5) = -24a^{15}b^9$.

✎ Multiply the following expressions.

19) $3x^2 \times 8x^3 =$

20) $2x^4 \times 9x^3 =$

21) $-4a^4b \times 2ab^3 =$

22) $(-7x^3yz) \times (3xy^2z^4) =$

23) $-2a^5bc \times 6a^2b^4 =$

24) $9u^3t^2 \times (-2ut) =$

25) $12x^2z \times 3xy^3 =$

26) $11x^3z \times 5xy^5 =$

27) $-6a^3bc \times 5a^4b^3 =$

28) $-4x^6y^2 \times (-12xy)$

Dividing Monomials

When dividing monomials, divide the numerical coefficients and divide the exponents with the same variable each other. For example, $6x^2 \div 2x = 3x$. It's important to make sure that the divisor and dividend have the same base before dividing.

It's also important to note that when dividing monomials, we can also use fractions. For example, $6x^2 \div 2x = \frac{6x^2}{2x} = 3x$.

If you divide a monomial with an exponent by another monomial with the same base, you subtract the exponent of the divisor from the exponent of the dividend. For example, $\frac{8x^3}{2x^2} = 4x^{(3-2)} = 4x$.

Summarily, keep in mind the following when working with monomials:
- When you divide two monomials, you need to divide their coefficients and then divide their variables.
- In the case of exponents with the same base, for Division, subtract their powers, for Multiplication, add their powers.
- Exponent's Division rules:

$$\frac{x^a}{x^b} = x^{a-b}$$

Here are a few samples:

Divide $18x^4$ by $6x^2$.

To divide these two monomials, we first make sure that they have the same base, which in this case is x. Then, we divide the numerical coefficients (18 and 6). So, $18 \div 6 = 3$. Next, divide the variables (x) with their exponents (4 and 2):

$$x^4 \div x^2 = x^{(4-2)} = x^2.$$

The result is, $(18x^4) \div (6x^2) = 3x^2$.

 Divide $\frac{12y^3}{-3y^4}$.

EffortlessMath.com

To divide these two monomials, we first make sure that they have the same base, which in this case is y. Then we divide the numerical coefficients (12 and -3), which is $12 \div (-3) = -4$, and divide the variables (y) with their exponents (3 and 4) as $y^3 \div y^4 = y^{(3-4)}$. The result is $\frac{12y^3}{-3y^4} = -4y^{(3-4)} = -4y^{-1} = -\frac{4}{y}$.

Divide expressions: $\frac{12x^4y^6}{6xy^2}$.

Use the division property of exponents:

$\frac{x^a}{x^b} = x^{a-b}$.

Divide the exponent with the same base:

$\frac{x^4}{x} = x^{4-1} = x^3$, and $\frac{y^6}{y^2} = y^{6-2} = y^4$.

Now, substitute the obtained expression in the original expression. Then: $\frac{12x^4y^6}{6xy^2} = 2x^3y^4$.

Divide the expressions: $(3a^5) \div \left(\frac{1}{9}a^4\right)$.

First, divide the numerical coefficient: $3 \div \frac{1}{9} = 3 \times 9 = 27$.

Next, use the multiplication property of exponents:

$\frac{x^a}{x^b} = x^{a-b} \to a^5 \div a^4 = a^1 = a$. Then: $(3a^5) \div \left(\frac{1}{9}a^4\right) = 27a$.

Divide expressions: $\frac{49a^6b^9}{7a^3b^4}$

Use the division property of exponents:

$\frac{x^a}{x^b} = x^{a-b} \to \frac{a^6}{a^3} = a^{6-3} = a^3$ and $\frac{b^9}{b^4} = b^{9-4} = b^5$.

Then: $\frac{49a^6b^9}{7a^3b^4} = 7a^3b^5$.

✎ Divide the following expressions.

29) $\frac{42x^4y^2}{6x^3y} =$

30) $\frac{49x^5y^6}{7x^2y} =$

31) $\frac{63x^{15}y^{10}}{9x^8y^6} =$

32) $\frac{35x^8y^{12}}{5x^4y^8} =$

Multiplying a Polynomial and a Monomial

To multiply a polynomial and a monomial, you simply multiply each term of the polynomial by the monomial. For example, if the polynomial is $2x^2 + 3x - 4$ and the monomial is $5x$, the product would be:

$(2x^2 + 3x - 4)(5x) = 10x^3 + 15x^2 - 20x$.

In general, the distributive property applies to any polynomial and monomial multiplication. The monomial $(5x)$ is distributed to every term in the polynomial, and the exponent of the variable in the monomial is added to the exponent of the variable in each term of the polynomial. So, when you multiply $2x^2$ and $5x$, the result is $(2x^2)(5x) = (2 \times 5)x^{2+1} = 10x^3$, similarly for other terms.

Notice that the order of the terms in the polynomial doesn't change during this operation, the result is still a polynomial.

Here are some examples of polynomial and monomial multiplication problems with detailed solutions:

Multiply $(3x^2 - 2x + 5)$ and $(3x)$.

By using the distributive property, we have:

$(3x^2 - 2x + 5)(3x) = (3x^2 \times 3x) - (2x \times 3x) + (5 \times 3x)$

Calculate the product of each of the monomial multiplications inside the parentheses:

$3x^2 \times 3x = (3 \times 3)x^{2+1} = 9x^3$,

$2x \times 3x = (2 \times 3)x^{1+1} = 6x^2$,

$5 \times 3x = (5 \times 3)x = 15x$

Put these expressions:

$(3x^2 - 2x + 5)(3x) = 9x^3 - 6x^2 + 15x$

Multiply $(2x^3 - 4x^2 + 5x + 3)$ and $(3x^2)$.

Use the distributive property:

$(2x^3 - 4x^2 + 5x + 3)(3x^2) = (2x^3 \times 3x^2) - (4x^2 \times 3x^2)$
$$+(5x \times 3x^2) + (3 \times 3x^2)$$

Multiply the monomials in parentheses:

$2x^3 \times 3x^2 = 6x^5$, $4x^2 \times 3x^2 = 12x^4$, $5x \times 3x^2 = 15x^3$, and $3 \times 3x^2 = 9x^2$.

Next, substitute them:

$(2x^3 - 4x^2 + 5x + 3)(3x^2) = 6x^5 - 12x^4 + 15x^3 + 9x^2$.

Multiply $(6x^4 + x^2 - 7)$ and (6).

To multiply these two expressions, we use the distributive property:

$(6x^4 + x^2 - 7)(6) = (6x^4 \times 6) + (x^2 \times 6) - (7 \times 6) = 36x^4 + 6x^2 - 42$

As you can see in each example, the distributive property is used to distribute the monomial to each term of the polynomial, and the exponents of the variable in the monomial are added to the exponents of the variable in each term of the polynomial.

Keep in mind that in the last example, the monomial is a constant, in this case, the constant is multiplied by each term of the polynomial.

Also, pay attention to the following examples:

Multiply expressions: $6x(2x + 5)$.

Use Distributive Property:

$6x(2x + 5) = (6x \times 2x) + (6x \times 5) = 12x^2 + 30x$.

Multiply: $-x(-2x^2 + 4x + 5)$.

Use Distributive Property:

$-x(-2x^2 + 4x + 5) = (-x) \times (-2x^2) + (-x) \times (4x) + (-x) \times (5) = 2x^3 - 4x^2 - 5x$

✍ Multiply the following expressions.

33) $3x(5x - y) =$

34) $2x(4x + y) =$

35) $7x(x - 3y) =$

36) $x(2x^2 + 2x - 4) =$

37) $5x(3x^2 + 8x + 2) =$

38) $7x(2x^2 - 9x - 5) =$

bit.ly/3aBYdx2

Multiply Polynomials Using Area Models

To multiply polynomials using area models, you can follow these steps:

Step 1: Create a rectangular area with one side corresponding to each of the polynomials.

Step 2: Divide each side of the rectangle corresponding to the polynomials into the monomial factors.

Step 3: Multiply the monomials in each area to complete the rectangle.

Step 4: To find the product of the polynomials, add the resulting expressions.

Let's review some examples:

Use the area model to find the product $2x(x + 1)$.

Model a rectangular area,

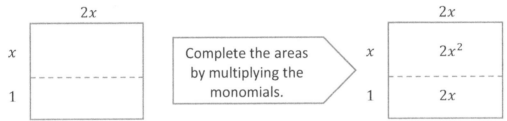

Last, combine terms to find the polynomial product.

$2x(x + 1) = 2x^2 + 2x$.

Use an area model to multiply these binomials.

$$(a - 2)(3a + 1)$$

Draw an area model representing the product $(a - 2)(3a + 1)$.

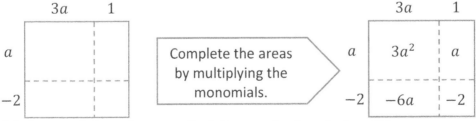

Now, add the partial products to find the product and simplify,

$3a^2 + a - 6a - 2 = 3a^2 - 5a - 2$.

Therefore, $(a - 2)(3a + 1) = 3a^2 - 5a - 2$.

Multiply $(3x^2 - 2x - 1)(1 - 3x)$, using the area model.

For this purpose, we model the multiplication as below.

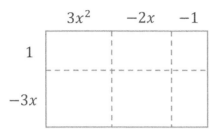

Multiply the monomials in each area to complete the rectangle.

	$3x^2$	$-2x$	-1
1	$3x^2$	$-2x$	-1
$-3x$	$-9x^3$	$6x^2$	$+3x$

Now, add the obtained expressions on the grid, to find the product and simplify.

Therefore:

$$(3x^2 - 2x - 1)(1 - 3x) = 3x^2 - 9x^3 - 2x + 6x^2 - 1 + 3x$$
$$= -9x^3 + (3x^2 + 6x^2) + (-2x + 3x) - 1$$
$$= -9x^3 + 9x^2 + x - 1$$

✍ Multiply the following expressions with the area model.

39) $3x(x + 2)$

40) $(a - 3)(2a + 2)$

✍ Complete the model area below and write the multiplication.

41)

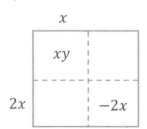

Multiplying Binomials

Multiplying binomials is a specific case of polynomial multiplication, where the polynomials are two terms polynomials, called binomials. The most common method of multiplying binomials is the FOIL method (First, Outer, Inner, Last). FOIL method is used to multiply two binomials of the form $(a + b)$ and $(c + d)$ as follows:

$$(a + b)(c + d) = ac + ad + bc + bd$$

It's based on distributing each term of the first binomial to each term of the second binomial. The acronym FOIL stands for:

- First: Multiply the first term of the first binomial by the first term of the second binomial.
- Outer: Multiply the first term of the first binomial by the second term of the second binomial.
- Inner: Multiply the second term of the first binomial by the first term of the second binomial.
- Last: Multiply the second term of the first binomial by the second term of the second binomial.

For example, to multiply $(x + 2)$ and $(x - 3)$, using the FOIL method:

$$(x + 2)(x - 3) = x(x) + x(-3) + 2(x) + 2(-3)$$
$$= x^2 - 3x + 2x - 6 = x^2 - x - 6$$

Another method to multiply binomials is using distributive property and combining like terms, but FOIL is the most widely used method.

It's important to notice that, when multiplying binomials, the order of terms in the binomials doesn't change, the result is still a polynomial.

Here's another example of multiplying binomials using FOIL method: Multiply $(3a - 2)$ and $(4a + 5)$.

EffortlessMath.com

Using the FOIL method:
$$(3a - 2)(4a + 5) = 3a(4a) + 3a(5) + (-2)(4a) + (-2)(5)$$
$$= 12a^2 + 15a - 8a - 10$$
$$= 12a^2 + 7a - 10$$

So, the product of $(3a - 2)$ and $(4a + 5)$ is $12a^2 + 7a - 10$.

Likewise, see some other examples:

Multiply Binomials. $(1 - x)(x - 2) =$

Use "FOIL". (First–Out–In–Last):
$$(1 - x)(x - 2) = x - 2 - x^2 + 2x.$$

Then combine like terms: $x - 2 - x^2 + 2x = -x^2 + 3x - 2$.

Therefore, the product of this multiplication is,
$$(1 - x)(x - 2) = -x^2 + 3x - 2.$$

Multiply Binomials. $(2a^3 - a)(a + 1) =$

Use "FOIL". (First–Out–In–Last):
$$(2a^3 - a)(a + 1) = 2a^4 + 2a^3 - a^2 - a.$$

So, the answer is this expression.

Multiply. $(x - 2y)(2x + 1) =$

Use "FOIL". (First–Out–In–Last):
$$(x - 2y)(2x + 1) = 2x^2 + x - 4xy - 2y.$$

Then simplify: $2x^2 + x - 4xy - 2y = 2x^2 - 4xy + x - 2y$.

✎ Multiply the following binomials.

42) $(x - 3)(x + 3) =$

43) $(x - 6)(x + 6) =$

44) $(x + 10)(x + 4) =$

45) $(x - 6)(x + 7) =$

46) $(x + 2)(x - 5) =$

47) $(x - 10)(x + 3) =$

Multiply two Binomials Using Algebra Tiles

To multiply two binomials using algebra tiles, you can follow these steps:

1. Place one binomial on the left side of the board and the other binomial on the right side of the board.
2. Take the first term of the left binomial and match it with the first term of the right binomial. This will create a rectangle tile with the product of the two terms.
3. Take the first term of the left binomial and match it with the second term of the right binomial. This will create a rectangle tile with the product of the two terms.
4. Take the second term of the left binomial and match it with the first term of the right binomial. This will create a rectangle tile with the product of the two terms.
5. Take the second term of the left binomial and match it with the second term of the right binomial. This will create a rectangle tile with the product of the two terms.
6. Arrange the rectangle tiles in a square array and add the products together to get the final result.

Pay attention an example for further explanation:

Use algebra tiles to simplify: $(x - 2)(2x + 1)$.

Set up the grid as follow:

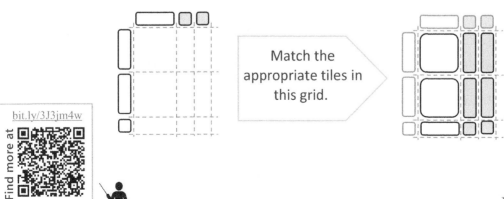

bit.ly/3J3jm4w

Find more at

Here, cross out one x tile on the first column and do the same on the second column.

Count the like terms inside the grid. Since the number of x^2 tiles = 2, the number of $-x$ tiles = 3, and the number of -1 tiles = 2, then, sum the like terms inside the grid. So, $2x^2 - 3x - 2$

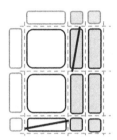

One more example:

Use algebra tiles to multiply $(2x - 1)$ to $(x - 3)$.

First, we set the binomial multiplication of a grading.

We see that there is no similar tile with the opposite sign in the grading.

Therefore, the number of tiles is obtained by multiplication.

Finally, we have:
$$(2x - 1)(x - 3) = 2x^2 - 7x + 3$$

✎ Multiply the following expressions by using the algebra tiles.

48) $(x + 1)(x + 6)$

49) $(2x + 1)(x - 4)$

Factoring Trinomials

Factoring trinomials is the process of expressing a trinomial, which is a polynomial with three terms, as the product of two binomials. One method for factoring trinomials is the "AC method", which involves identifying the coefficient of the squared term (a) and the constant term (c) and then finding two numbers whose product is c and whose sum is the coefficient of the x term (b). These two numbers can be used as the factors of the trinomial. Another method is known as "grouping" where you factor out the greatest common factor from the first two terms and the last two terms and then factor those two binomials. For example, to factor the trinomial $x^2 - 5x + 6$, we could use the AC method by identifying $a = 1$, $b = -5$, and $c = 6$. The two numbers whose product is 6 and whose sum is -5 are -2 and -3, so we can factor the trinomial as $(x - 2)(x - 3)$. Alternatively, we could use the grouping method. First, rewrite the expression as $x^2 - 3x - 2x + 6$. Next, we factor out the greatest common factor of x^2 and $3x$ which is x, this gives us $x(x - 3) - 2x + 6$. Now we can factor the remaining binomial $-2x + 6$ which gives us $x(x - 3) - 2(x - 3)$ and thus the original trinomial can be written as $x(x - 3) - 2(x - 3) = (x - 3)(x - 2)$.

Let's review two examples:

Factor $x^2 + 7x + 10$.

We can use the AC method to factor this trinomial. The coefficient of the squared term is $a = 1$, the coefficient of the x term is $b = 7$, and the constant term is $c = 10$. We can find two numbers whose product is 10 and whose sum is 7: 2 and 5. So, we can factor the trinomial as $(x + 2)(x + 5)$.

To factor trinomials, you can use the following methods:

- "FOIL": $(x + a)(x + b) = x^2 + (b + a)x + ab$.
  - "Difference of Squares":
    $$a^2 + 2ab + b^2 = (a + b)(a + b)$$

Polynomials

$$a^2 - 2ab + b^2 = (a-b)(a-b)$$

- "Reverse FOIL": $x^2 + (b+a)x + ab = (x+a)(x+b)$.

Some another examples:

Factor this trinomial. $x^2 - 2x - 8$

Break the expression into groups. You need to find two numbers that their product is -8 and their sum is -2.

(Remember "Reverse FOIL": $x^2 + (b+a)x + ab = (x+a)(x+b)$).

Those two numbers are 2 and -4. Then: $x^2 - 2x - 8 = (x^2 + 2x) + (-4x - 8)$.

Now factor out x from $x^2 + 2x$: $x(x+2)$, and factor out -4 from $-4x - 8$: $-4(x+2)$; Then: $(x^2 + 2x) + (-4x - 8) = x(x+2) - 4(x+2)$.

Now factor out like term: $(x+2)$. Then: $(x+2)(x-4)$.

Factor this trinomial. $4x^2 + 4x - 8$

First, factor -4 from the trinomial and rewrite as $4(x^2 + x - 2)$. Next, use the AC method for $x^2 + x - 2$. Find two numbers whose product is -2 and whose sum is 1 are 2 and -1, so we can factor the trinomial. Therefore,

$$4x^2 + 4x - 8 = 4(x-1)(x+2).$$

Factor this trinomial. $4x^6 - 10x^3 + 6$

Comparing the given trinomial and the "Reverse FOIL" method

$$x^2 + (b+a)x + ab = (x+a)(x+b),$$

we find that by substituting $2x^3$ for x and rewriting the trinomial as we get: $(2x^3)^2 - 5(2x^3) + 6$. Next, use the "AC method" for resulting trinomial, we have: $4x^6 - 10x^3 + 6 = (2x^3 - 2)(2x^3 - 3)$.

✏️ Factor the following expressions.

50) $x^2 + 6x + 8 =$

51) $x^2 + 3x - 10$

52) $x^2 + 2x - 48 =$

53) $x^2 - 10x + 16 =$

54) $2x^2 - 10x + 12 =$

55) $3x^2 - 10x + 3 =$

Factoring Polynomials

Factoring polynomials is the process of expressing a polynomial as the product of simpler polynomials. There are several methods for factoring polynomials, including:

1. Factoring by grouping: This method involves grouping the terms of the polynomial into pairs and factoring out the greatest common factor from each pair.

2. Factoring by using the difference of squares: This method involves identifying two binomials that when multiplied together give the polynomial, and are each the square of the difference of two terms.

3. Factoring by using the sum and difference of cubes: This method involves identifying two binomials that when multiplied together give the polynomial, and are each the cube of the sum or difference of two terms.

4. Factoring by using the Rational Root Theorem: This method involves finding the possible rational roots of the polynomial, and testing them to see if they divide the polynomial.

5. Factoring by synthetic division: This method involves dividing the polynomial by a linear factor using synthetic division.

For example, to factor the polynomial $x^3 + 6x^2 - x - 6$, we could use factoring by grouping. We would group the first two terms, $x^3 + 6x^2$, and the last two terms, $-x - 6$, and find that both pairs share a common factor of $x + 6$. So, we can factor out $x + 6$ and get:

$$x^3 + 6x^2 - x - 6 = (x^3 + 6x^2) - x - 6 = x^2(x + 6) - x - 6$$
$$= x^2(x + 6) - (x + 6) = (x + 6)(x^2 - 1).$$

EffortlessMath.com

Polynomials

Now, we can factor the remaining binomial $x^2 - 1$ as $(x-1)(x+1)$, and the original polynomial can be written as $(x+6)(x+1)(x-1)$.

Another example is to factor the polynomial $x^4 - 16$, we can use the difference of squares method by noticing that $x^4 - 16 = (x^2 + 4)(x^2 - 4)$.

It's worth noting that not all polynomials can be factored and the method of factoring will depend on the polynomial you are trying to factor.

Some more examples:

Factor each polynomial.

$$6x^3 - 6x$$

To factorize the expression $6x^3 - 6x$, first factor out the largest common factor, $6x$, and then you will see that you have the pattern of the difference between two complete squares: $(a-b)(a+b) = a^2 - b^2$,

then: $6x^3 - 6x = 6x(x^2 - 1) = 6x(x-1)(x+1)$.

$$x^2 + 9x + 20$$

To factorize the expression $x^2 + 9x + 20$, you need to find two numbers whose product is 20 and sum is 9. You can get the number 20 by multiplying $1 \times 20, 2 \times 10, 4 \times 5$. The last pair will be your choice, because $4 + 5 = 9$.

Then: $x^2 + 9x + 20 = (x+4)(x+5)$.

✎ Factor each expression.

56) $4x^2 - 4x - 8$

57) $6x^2 + 37x + 6$

58) $16x^2 + 60x - 100$

59) $4x^2 - 17x + 4$

Use a Graph to Factor Polynomials

If a polynomial includes the factor of the form $(x - h)^p$, you can determine the behavior near the x-intercept by the power p. It can be said that $x = h$ is a zero of multiplicity p:

- If a polynomial function graph touches the x-axis, it's a zero with even multiplicity.
- If a polynomial function graph crosses the x-axis, it's a zero with odd multiplicity.
- A polynomial function graph gets flattered at zero if the multiplicity of the zero is higher.
- The sum of the multiplicities of the zero is the polynomial function's degree.

To check factorization using a graphing calculator, follow these steps:

Step 1: Press the $Y =$ button and enter the given equation for $Y1$.

Step 2: Press the GRAPH button to see the equation's graph.

Step 3: Press the TRACE button and by the left and right buttons move the cursor along the graph. You can see at which points the graph crosses the x-axis.

Step 4: To find the y-values at the x-values when the graph crosses the x-axis, enter the value of x at this point and press the ENTER button while in Trace mode. The calculator finds the y-value for you. The calculator tells you that in these values of x, the y-values are equal to zero.

Step 5: Remember that for functions with binomial factors of the form $(x - a)$, a is an x-intercept.

Let's review an example:

Use a graph to factor following polynomial: $y = x^2 - x - 2$

First, graph the polynomial. Then find the points where the polynomial function graph crosses the x-axis. These points are the zeros of the polynomial function. For $y = x^2 - x - 2$, $x = -1$ and $x = 2$ are the zeros of the polynomial function.

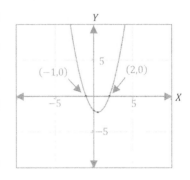

Therefore, two binomials $(x + 1)$ and $(x - 2)$ are the factors of the given trinomial. So, we can write as follows:
$$x^2 - x - 2 = (x + 1)(x - 2).$$
Use the following graph to find the polynomial factors below:
$$y = 2x^4 - 7x^3 + 6x^2 + x - 2$$
To solve, we first find the zeros of the polynomial from the graph. We can see that the graph intersects the x-axis at points $-\frac{1}{2}$ and 2 (with odd multiplicity) and touches at point 1 (with even multiplicity).

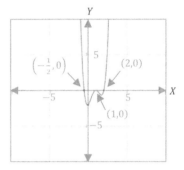

So, you can write the given polynomial in the form:
$$y = Q(x)\left(x + \frac{1}{2}\right)^r (x - 1)^s (x - 2)^t,$$
where r and t are odd positive integers and s is an even positive integer, and $r + s + t \le 4$. $Q(x)$ is a non-zero polynomial whose power of the leading term is $4 - (r + s + t)$. Clearly, we have: $r = 1$, $s = 2$, and $t = 1$. Then, Q is a constant number such as k. Because every polynomial with zero degree is a constant number. So, we get: $y = k\left(x + \frac{1}{2}\right)(x - 1)^2(x - 2)$.

To find k, the factors can be multiplied together and compared with the initial polynomial. $y = 2\left(x + \frac{1}{2}\right)(x - 1)^2(x - 2)$.

🖎 Use a graph to find zeros of following polynomials.

60) $y = x^2 - 4$

61) $y = -(x + 2)^2$

Factoring Special Case Polynomials

There are several special cases of polynomials that can be factored using specific methods. These include:

1. Factoring perfect square trinomials: A trinomial of the form $a^2 + 2ab + b^2$ or $a^2 - 2ab + b^2$ is a perfect square trinomial, which can be factored as $(a + b)^2$ or $(a - b)^2$ respectively.

2. Factoring a difference of squares: A polynomial of the form $a^2 - b^2$ can be factored as $(a + b)(a - b)$ using the difference of squares identity.

3. Factoring a sum of cubes: A polynomial of the form $a^3 + b^3$ can be factored as $(a + b)(a^2 - ab + b^2)$ using the sum of cubes identity.

4. Factoring a difference of cubes: A polynomial of the form $a^3 - b^3$ can be factored as $(a - b)(a^2 + ab + b^2)$ using the difference of cubes identity.

5. Factoring by grouping: A polynomial of the form $ax^3 + bx^2 + cx + d$ can be factored by grouping the first two terms and the last two terms and factoring out the greatest common factor from each pair.

6. Factoring by synthetic division: You can use synthetic division to divide the polynomial by a linear factor.

For example, to factor the trinomial $x^2 - 4x + 4$, we can see that it's a perfect square trinomial and we can factor it as $(x - 2)^2$.

Another example: to factor the polynomial $x^3 + 8$, we can factor it as $(x + 2)(x^2 - 2x + 4)$ by using the difference of cubes identity.

It's important to note that not all polynomials will have special cases that can be used to factor them, and it's not always the most efficient method. It's also important to check the solutions of the polynomial equation before applying the factoring methods because that can make the process easier.

Here are some more examples for further explanation:

Factor completely: $x^3 - x^2 - 9x + 9$.

Looking at the given equation, we can see that the factor x^2 can be factored from two monomials with the largest degree $(x^3 - x^2)$. Thus:

$x^3 - x^2 - 9x + 9 = x^2(x - 1) - 9x + 9$.

Also, we can see factor from the coefficient -9 of this part of the polynomial $-9x + 9$ to get binomial $(x - 1)$. So, we have:

$x^2(x - 1) - 9x + 9 = x^2(x - 1) - 9(x - 1)$.

In this step, we separate factor $(x - 1)$ as a coefficient from the whole polynomial.

$x^2(x - 1) - 9(x - 1) = (x - 1)(x^2 - 9)$

Looking more closely at the result, we can see that the monomial $(x^2 - 9)$ can be factored using the formula $a^2 - b^2 = (a + b)(a - b)$. Therefore, we get:

$x^2 - 9 = (x + 3)(x - 3)$.

Finally, substitute the obtained polynomial in the original expression to get the answer: $x^3 - x^2 - 9x + 9 = (x - 1)(x + 3)(x - 3)$.

Factor: $36y^2 - 25x^4$.

Phrase $36y^2$ can be written in form $(6y)^2$ and phrase $25x^4$ in form $(5x^2)^2$. Now, according to the formula $a^2 - b^2 = (a + b)(a - b)$. Put the values of $a = 6y$ and $b = 5x^2$.

Therefore, the relation of the question form is as follows:

$(6y)^2 - (5x^2)^2 = (6y - 5x^2)(6y + 5x^2)$.

✎ Factor each completely.

62) $36x^2 - 121 =$

63) $-36x^4 + 4x^2 =$

64) $-36x^2 + 400 =$

65) $49x^2 - 56x + 16 =$

66) $1 - x^2 =$

67) $81x^4 - 900x^2 =$

Add Polynomials to Find Perimeter

When working with polynomials, adding them is a process of combining like terms. This can be done by adding the coefficients of the same degree together. To add polynomials to find the perimeter of a shape, you will need to first identify the polynomials that represent the lengths of the sides of the shape.

For example, let's say we have a rectangle with sides of length $x + 2$ and $x - 3$. To find the perimeter of this rectangle, we would add the polynomials representing the lengths of the sides:

$2((x + 2) + (x - 3)) = 2(2x - 1) = 4x - 2$.

So, the perimeter of the rectangle is represented by the polynomial $4x - 2$.

For another example, let's say we have a square with side length $2x + 3$. To find the perimeter of this square, we would add the polynomials representing the lengths of the sides: $4(2x + 3) = 8x + 12$.

So, the perimeter of the square is represented by the polynomial $8x + 12$.

It's worth noting that the perimeter is a scalar value, so the polynomial representing it will be a constant, and it's not dependent on the variable.

Also, the same process can be used for different shapes and different dimensions, you just need to identify the polynomials that represent the lengths of the sides and then add them together to find the perimeter.

Here's another example:

Find the perimeter. Simplify your answer.

The perimeter of the shape is the sum of the sides. So,

Perimeter $= (x - 2) + (2x) + (3x - 1) + (2x)$

$= x - 2 + 2x + 3x - 1 + 2x$.

Group and add like terms,

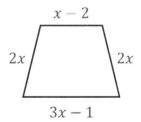

Polynomials 199

Perimeter $= (x + 2x + 3x + 2x) + (-2 - 1) = 8x - 3$.

What is the perimeter of the rectangle if the length is $2x^2 - 1$ and the width is $4(x + 1)$?

The perimeter of the rectangle is,

$$\text{Perimeter} = 2((2x^2 - 1) + 4(x + 1))$$

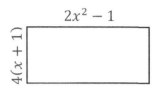

Expand the expression and simplify,

$$\text{Perimeter} = 2((2x^2 - 1) + (4x + 4))$$
$$= 2(2x^2 - 1 + 4x + 4)$$
$$= 2(2x^2 + 4x + 3)$$
$$= 4x^2 + 8x + 6.$$

🖎 What is the perimeter of the following shapes.

68) Perimeter =

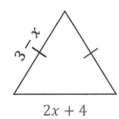

69) Perimeter =

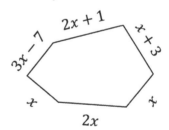

🖎 Based on the information in each case, find the value of x.

70) $a =$ _____

Perimeter $= 36$

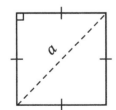

71) $x =$ _____

Perimeter $= 42$

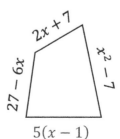

Chapter 7: Answers

1) $18x + 12$

2) $15x - 40$

3) $7x^2 + 11x$

4) $6x^2 + 23x$

5) $18x^2 + x$

6) $6x^2 - 4x - 6$

7) $-2x^2 - 24x - 5$

8) $-10x^2 - 30x + 7$

17) $x^2 - 2x + 4$

19) $24x^5$

20) $18x^7$

21) $-8a^5b^4$

22) $-21x^4y^3z^5$

23) $-12a^7b^5c$

24) $-18u^4t^3$

25) $36x^3y^3z$

26) $55x^4y^5z$

27) $-30a^7b^4c$

28) $48x^7y^3$

9) $3x^2 - 1$

10) $2x^2 - 14x$

11) $6x^3 - 8x^2$

12) $x^3 - 4x$

13) $10x^3 + 18x^2 - 8$

14) $x^3 + 7x^2 - 9$

15) $3x^3 + 7x$

16) $4x^3 - 10x$

18) $3x^2 + 4x + 6$

29) $7xy$

30) $7x^3y^5$

31) $7x^7y^4$

32) $7x^4y^4$

33) $15x^2 - 3xy$

34) $8x^2 + 2xy$

35) $7x^2 - 21xy$

36) $2x^3 + 2x^2 - 4x$

37) $15x^3 + 40x^2 + 10x$

38) $14x^3 - 63x^2 - 35x$

Polynomials

39) $3x^2 + 6x$

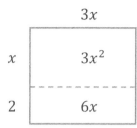

40) $2a^2 - 4a - 6$

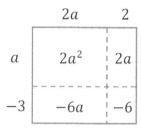

41) $(x-1)(y+2x) = xy + 2x^2 - y - 2x$

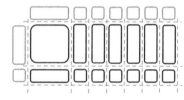

42) $x^2 - 9$

43) $x^2 - 36$

44) $x^2 + 14x + 40$

45) $x^2 + x - 42$

46) $x^2 - 3x - 10$

47) $x^2 - 7x - 30$

48) $x^2 + 7x + 6$

49) $2x^2 - 7x - 4$

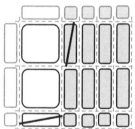

50) $(x+4)(x+2)$

51) $(x+5)(x-2)$

52) $(x-6)(x+8)$

53) $(x-8)(x-2)$

54) $2(x-2)(x-3)$

55) $(3x-1)(x-3)$

56) $4(x+1)(x-2)$

57) $(x+6)(6x+1)$

58) $4(x+5)(4x-5)$

59) $(x-4)(4x-1)$

60) $x = \pm 2$

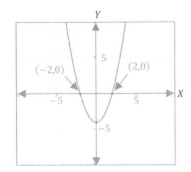

61) $x = -2$

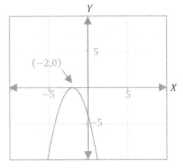

62) $(6x - 11)(6x + 11)$

63) $-4x^2(3x + 1)(3x - 1)$

64) $-4(3x + 10)(3x - 10)$

65) $(7x - 4)^2$

66) $(1 + x)(1 - x)$

67) $9x^2(3x + 10)(3x - 10)$

68) 10

69) $10x - 3$

70) $a = 9\sqrt{2}$

71) $x = 4$

CHAPTER
8 Relations and Functions

Math topics that you'll learn in this chapter:

- ☑ Function Notation and Evaluation
- ☑ Adding and Subtracting Functions
- ☑ Multiplying and Dividing Functions
- ☑ Composition of Functions
- ☑ Evaluate an Exponential Function
- ☑ Function Inverses
- ☑ Domain and Range of Relations
- ☑ Interval Notation
- ☑ Rate of Change and Slope
- ☑ Complete a Function Table from an Equation

Function Notation and Evaluation

Function notation is a way to express a mathematical function using a specific symbol, usually denoted by "$f(x)$" or "$g(x)$" where "x" is the input (also called the independent variable) and "$f(x)$" or "$g(x)$" is the output (also called the dependent variable). The notation is read as "f of x" or "g of x", and it's used to indicate that the value of "$f(x)$" or "$g(x)$" is determined by the value of "x".

For example, if we have a function $f(x) = 2x + 3$, this means that for any value of x, the corresponding value of $f(x)$ is found by plugging that value of x into the equation and solving for the output. So, if we plug in $x = 2$, we would get:

$f(2) = 2(2) + 3 = 4 + 3 = 7$.

Evaluation of a function is the process of determining the value of the function at a specific input.

For example, if we have a function $g(x) = x^2$, and we want to find the value of $g(-1)$, we would substitute -1 for x in the equation and solve for the output: $g(-1) = (-1)^2 = 1$. So, the value of $g(-1)$ is 1.

It's worth noting that in the function notation, the letter "f" or "g" is arbitrary and can be replaced by any other letter.

Also, you can have a function with more than one variable. Like $f(x, y) = x^2 + y^2$, where x and y are the independent variable and $f(x, y)$ is the dependent variable.

Furthermore, you can have a function defined by a table of values, where you have the function inputs and their corresponding outputs, and the function can be defined by the relation between the inputs and the outputs.

Let's review an example:

Find the value of the function $f(x) = 2x^2 + 4$ for -1 and 1.

For this purpose, plug the value of $x = -1$ into the function $f(x) = 2x^2 + 4$, as $f(-1) = 2(-1)^2 + 4$. Then, calculate the value of expression:

$2(-1)^2 + 4 = 2(1) + 4 = 2 + 4 = 6$.

Here, I see that the value of the function $f(x)$ for -1 is 6.

Similarly, we find the value of the function for 1.

$x = 1 \rightarrow f(1) = 2(1)^2 + 4 = 2(1) + 4 = 2 + 4 = 6$.

We see that the function values for -1 and 1 are equal to each other.

Look at some other examples:

Evaluate: $f(x) = x + 6$, find $f(2)$.

Substitute x with 2. Then:

$f(x) = x + 6 \rightarrow f(2) = 2 + 6 \rightarrow f(2) = 8$.

Therefore, the corresponding value of $f(x) = x + 6$ for $x = 2$ is 8.

Evaluate: $w(x) = 3x - 1$, find $w(4)$.

Substitute x with 4. Then: $w(x) = 3x - 1 \rightarrow w(4) = 3(4) - 1 = 12 - 1 = 11$.

So, the value of $w(4)$ is 11.

Evaluate: $h(x) = 4x^2 - 9$, find $h(2a)$.

Substitute x with $2a$. Then:

$h(x) = 4x^2 - 9 \rightarrow h(2a) = 4(2a)^2 - 9 \rightarrow h(2a) = 4(4a^2) - 9 = 16a^2 - 9$.

Here, the value of the function $h(x)$ is a combination of the value $2a$ evaluated the values of h by putting different values for a.

✎ Evaluate each function.

1) $f(x) = x - 2$, find $f(-1)$

2) $g(x) = 2x + 4$, find $g(3)$

3) $g(n) = 2n - 8$, find $g(-1)$

4) $h(n) = n^2 - 1$, find $h(-2)$

5) $f(x) = x^2 + 12$, find $f(5)$

6) $g(x) = 2x^2 - 9$, find $g(-2)$

7) $w(x) = 2x^2 - 4x$, find $w(2n)$

8) $p(x) = 4x^3 - 10$, find $p(-3a)$

Adding and Subtracting Functions

When adding or subtracting functions, we combine the expressions for each function and simplify the result.

To add two functions, we simply add their expressions together. For example, if we have two functions $f(x) = 2x + 3$ and $g(x) = x - 4$, we can add them together to get:

$f(x) + g(x) = (2x + 3) + (x - 4) = 2x + 3 + x - 4 = (2x + x) + (3 - 4) = 3x - 1$

The addition of two functions f and g is represented by the notation $f + g$.

$$(f + g)(x) = f(x) + g(x)$$

To subtract two functions, we subtract the expressions for one function from the other. For example, if we have two functions $f(x) = 2x + 3$ and $g(x) = x - 4$, we can subtract them to get:

$f(x) - g(x) = (2x + 3) - (x - 4) = 2x + 3 - x + 4 = (2x - x) + (3 + 4) = x + 7$

The subtraction of two functions f and g is represented by the notation $f - g$.

$$(f - g)(x) = f(x) - g(x)$$

It's worth noting that the same process can be applied to functions with more than one variable, you just need to combine the expressions of the functions and simplify the result.

Also, when you are working with different functions, you need to make sure that the domain of the function and the range of the function are the same, otherwise, the operation is not defined.

Additionally, if you want to add or subtract functions with different variables, you can use the principle of the change of variables.

For example, if you want to add $f(x) = 2x + 3$ and $g(y) = y - 4$, you can change the variable of $g(y)$ to $g(x) = x - 4$, and then add the two functions.

Some examples:

Let $g(x) = 2x - 2$, and $f(x) = x + 1$. Find: $(g + f)(x)$.

Relations and Functions

In order to, use the relation: $(f + g)(x) = f(x) + g(x)$. Then:
$(g + f)(x) = g(x) + f(x) = (2x - 2) + (x + 1) = 2x - 2 + x + 1$
$= (2x + x) + (-2 + 1) = 3x - 1.$

Therefore, the equation of the function $g + f$ is $(g + f)(x) = 3x - 1$.

Find $(f - g)(x)$, for the functions $f(x) = 4x - 3$, and $g(x) = 2x - 4$.

Given that $(f - g)(x) = f(x) - g(x)$. Then:

$(f - g)(x) = (4x - 3) - (2x - 4) = 4x - 3 - 2x + 4 = 2x + 1.$

If $h(x) = x^2 + 2$, and $k(x) = x + 5$. Find: $(h + k)(x)$.

According to the $(g + f)(x) = g(x) + f(x)$, we have:

$(h + k)(x) = h(x) + k(x) = (x^2 + 2) + (x + 5) = x^2 + x + 7.$

Evaluate: $f(x) = 5x^2 - 3$, $g(x) = 3x + 6$. Find: $(f - g)(3)$.

Then: $(f - g)(x) = (5x^2 - 3) - (3x + 6) = 5x^2 - 3 - 3x - 6 = 5x^2 - 3x - 9.$

Substitute x with 3: $(f - g)(3) = 5(3)^2 - 3(3) - 9 = 45 - 9 - 9 = 27.$

$g(x) = x^2 - 4$, $f(x) = 2x + 3$, Find: $(g + f)(x)$. Since $(g + f)(x) = g(x) + f(x)$.

Then: $(g + f)(x) = (x^2 - 4) + (2x + 3) = x^2 - 4 + 2x + 3 = x^2 + 2x - 1.$

✏ Perform the indicated operation.

9) $g(x) = x - 2,$

   $h(x) = 2x + 6$

   Find: $(h + g)(3)$

10) $f(x) = 3x + 2,$

    $g(x) = -x - 6$

    Find: $(f + g)(2)$

11) $f(x) = 5x + 8,$

    $g(x) = 3x - 12$

    Find: $(f - g)(-2)$

12) $h(x) = 2x^2 - 10.$

    $g(x) = 3x + 12$

    Find: $(h + g)(3)$

13) $g(x) = 12x - 8,$

    $h(x) = 3x^2 + 14$

    Find: $(h - g)(x)$

14) $h(x) = -2x^2 - 18,$

    $g(x) = 4x^2 + 15$

    Find: $(h - g)(a)$

Multiplying and Dividing Functions

When multiplying or dividing functions, we use the standard rules of algebra to simplify the result.

To multiply two functions, we simply multiply their expressions together. For example, if we have two functions $f(x) = 2x + 3$ and $g(x) = x - 4$, we can multiply them together to get:

$f(x) \cdot g(x) = (2x + 3) \cdot (x - 4) = 2x^2 - 8x + 3x - 12 = 2x^2 - 5x - 2$.

The multiplication of two functions f and g is represented by the notation $f \cdot g$.

$$(f \cdot g)(x) = f(x) \cdot g(x)$$

To divide two functions, we divide the expression of one function by the other. For example, if we have two functions $f(x) = 2x + 3$ and $g(x) = x - 4$, we can divide them to get: $\left(\frac{f}{g}\right)(x) = \frac{f(x)}{g(x)} = \frac{2x+3}{x-4}$.

The division of two functions as g by f is represented by the notation $\frac{f}{g}$.

$$\left(\frac{f}{g}\right)(x) = \frac{f(x)}{g(x)}$$

It's worth noting that when you divide two functions, you should check that the denominator is not equal to zero since it would mean that the division is not defined.

It's also worth noting that the same process can be applied to functions with more than one variable, you just need to combine the expressions of the functions and simplify the result.

Also, when you are working with different functions, you need to make sure that the domain of the function and the range of the function are the same, otherwise, the operation is not defined.

Additionally, if you want to multiply or divide functions with different variables, you can use the principle of the change of variables. For example, if

Relations and Functions

you want to multiply $f(x) = 2x + 3$ and $g(y) = y - 4$, you can change the variable of $g(y)$ to $g(x) = x - 4$, and then multiply the two functions.

Let's review some examples:

Let's $g(x) = x + 3$, and $f(x) = x + 4$. Find: $(g.f)(x)$.

Use the formula: $(f \cdot g)(x) = f(x) \cdot g(x)$. Instead of $f(x)$ and $g(x)$, put $g(x) = x + 3$ and $f(x) = x + 4$, respectively. So, we get:

$(g.f)(x) = g(x).f(x) = (x + 3)(x + 4) = x^2 + 4x + 3x + 12 = x^2 + 7x + 12$.

Evaluate: $f(x) = x + 6$, and $h(x) = x - 9$. Find: $\left(\frac{f}{h}\right)(x)$.

Since $\left(\frac{f}{g}\right)(x) = \frac{f(x)}{g(x)}$. Therefore, $\left(\frac{f}{h}\right)(x) = \frac{x+6}{x-9}$.

If $g(x) = x + 7$, and $f(x) = x - 3$. Find: $(g.f)(2)$.

To solve it, we can get the equation of function $(g.f)(x)$. Therefore, $(g.f)(x) = g(x).f(x) = (x + 7)(x - 3) = x^2 - 3x + 7x - 21$.

Then: $g(x).f(x) = x^2 + 4x - 21$.

Next, substitute x with 2: $(g.f)(2) = (2)^2 + 4(2) - 21 = 4 + 8 - 21 = -9$.

$f(x) = x + 3$, $h(x) = 2x - 4$. Find: $\left(\frac{f}{h}\right)(3)$.

Since $\left(\frac{f}{h}\right)(x) = \frac{f(x)}{h(x)} = \frac{x+3}{2x-4}$. Substitute x with 3: $\left(\frac{f}{h}\right)(3) = \frac{3+3}{2(3)-4} = \frac{6}{2} = 3$.

✎ Perform the indicated operation.

15) $g(x) = x - 5$, $h(x) = x + 6$

   Find: $(g.h)(-1) =$

16) $f(x) = 2x + 2$, $g(x) = -x - 6$

   Find: $\left(\frac{f}{g}\right)(-2) =$

17) $f(x) = 5x + 3$, $g(x) = 2x - 4$

   Find: $\left(\frac{f}{g}\right)(5) =$

18) $h(x) = x^2 - 2$, $g(x) = x + 4$

   Find: $(g.h)(3) =$

19) $g(x) = 4x - 12$, $h(x) = x^2 + 4$

   Find: $(g.h)(-2) =$

20) $h(x) = 3x^2 - 8$, $g(x) = 4x + 6$

   Find: $\left(\frac{h}{g}\right)(-4) =$

Composition of Functions

The composition of functions is a way to combine two or more functions to create a new function. It is denoted by the symbol $fog(x)$, and it is read as "f composed with g".

The composition of two functions $f(x)$ and $g(x)$ is defined as the function $h(x) = f(g(x))$, where $h(x)$ is the new function formed by the composition. This means that the output of function $g(x)$ is used as the input to function $f(x)$, and the final output of the composition is $h(x)$. Thus, it can be said:

$$fog(x) = f(g(x))$$

For example, if we have two functions $f(x) = 2x + 3$ and $g(x) = x^2$, we can compose them together to get: $h(x) = f(g(x)) = f(x^2) = 2(x^2) + 3 = 2x^2 + 3$.

It's worth noting that when you compose two functions, the domain of the composed function is the intersection of the domain of the two functions being composed. The range of the composed function is the range of the first function. Also, you can compose multiple functions together.

For example, if you have three functions: $f(x) = 2x + 3$, $g(x) = x^2$, and $h(x) = x^3$, you can compose them together in the order of $hogof(x) = h\big(g(f(x))\big)$. So, the output of $f(x)$ would be the input of $g(x)$, the output of $g(x)$ would be the input of $h(x)$, and the final output is the composition of all the functions.

Finally, it is worth noting that the composition of functions is not commutative, meaning that $f(g(x))$ is not the same as $g(f(x))$, and it is not associative meaning that $(fog)oh(x)$ is not the same as $fo(goh)(x)$.

Additionally, it is important to note that when composing functions, the order of the functions matters. The composition of two functions $f(x)$ and $g(x)$ is not the same as the composition of the same two functions in reverse order, i.e., $g(f(x))$. This is because the output of one function is used as the input of the other function, so the order in which they are composed affects the final output. For example, if we have two functions $f(x) = 2x + 3$, and $g(x) = x^2$, then the composition of $f(g(x)) = f(x^2) = 2x^2 + 3$, while the composition of $g(f(x)) = g(2x + 3) = (2x + 3)^2 = 4x^2 + 12x + 9$.

Let's review an example:

Using $f(x) = x + 3$ and $g(x) = 5x$, find: $(fog)(x)$.

According to $(fog)(x) = f(g(x))$. Then: $(fog)(x) = f(g(x)) = f(5x)$.

Now, find $f(5x)$ by substituting x with $5x$ in the $f(x)$ function.

Therefore, $f(x) = x + 3$; $(x \to 5x) \to f(5x) = (5x) + 3 = 5x + 3$.

Another example:

Using $f(x) = 3x - 1$ and $g(x) = 2x - 2$, find: $(gof)(5)$.

Using $(fog)(x) = f(g(x))$. Then: $(gof)(x) = g(f(x)) = g(3x - 1)$,

Now, substitute x in $g(x)$ by $(3x - 1)$.

Then: $g(3x - 1) = 2(3x - 1) - 2 = 6x - 2 - 2 = 6x - 4$.

Substitute x with 5: $(gof)(5) = g(f(5)) = 6(5) - 4 = 30 - 4 = 26$.

✎ Solve.

21) $f(x) = 2x$, $g(x) = x + 3$

Find: $(fog)(2) =$

22) $f(x) = x + 2$, $g(x) = x - 6$

Find: $(fog)(-1) =$

23) $f(x) = 3x$, $g(x) = x + 4$

Find: $(gof)(4) =$

24) $h(x) = 2x - 2$, $g(x) = x + 4$

Find: $(goh)(2) =$

25) $f(x) = 2x - 8$, $g(x) = x + 10$

Find: $(fog)(-2) =$

26) $f(x) = x^2 - 8$, $g(x) = 2x + 3$

Find: $(gof)(4) =$

Evaluate an Exponential Function

For any real number x, an exponential function is an equation with the form $f(x) = ab^x$, where a is a non-zero real number and b is a positive real number. For example, $f(x) = 2^x$, $g(x) = -3\left(\frac{1}{2}\right)^x$, and $h(x) = 5\left(\frac{7}{2}\right)^x$.

To evaluate the value of an exponential function, it's enough to substitute the given input for the independent variable. For example, the value of the function $f(x) = \frac{1}{2}(3)^x$ for the value of $x = 0$ is $f(0) = \frac{1}{2}$.

It is worth noting that exponential functions can be added and subtracted or combined with other functions.

Pay attention to the following explanations:

Find the value of the exponential function of $f(x) = 3^x$ for $x = 2$.

To solve, it is enough to substitute $x = 2$, in the equation $f(x) = 3^x$.

So, $f(2) = 3^2 = 9$.

Let $f(x) = 7(2)^x$. Evaluate $f(3)$.

To evaluate $f(3)$, plug 3 in equation $f(x) = 7(2)^x$ instead of the independent variable x. Therefore, $f(3) = 7(2)^3 = 7(8) = 56$.

Use the following function to find $f(6)$.

$$f(x) = -2(3)^{\frac{x}{2}-1}$$

First, substitute $x = 6$ in the equation,

$f(6) = -2(3)^{\frac{6}{2}-1} = -2(3)^{3-1} = -2(3)^2 = -2(9) = -18$.

Use the following function to find $f(2)$.

$$f(x) = 9\left(\frac{1}{3}\right)^{2x-1} + 1$$

To solve, plug 2 into $f(x) = 9\left(\frac{1}{3}\right)^{2x-1} + 1$ instead of x. So,

$f(2) = 9\left(\frac{1}{3}\right)^{2(2)-1} + 1 = 9\left(\frac{1}{3}\right)^{2(2)-1} + 1 = 9\left(\frac{1}{3}\right)^{4-1} + 1 = 9\left(\frac{1}{3}\right)^3 + 1$

Relations and Functions

$$= 9\left(\frac{1}{27}\right) + 1 = \frac{1}{3} + 1 = \frac{4}{3}.$$

Therefore, the value of $f(2)$ is $\frac{4}{3}$.

Another example:

Let $f(x) = 2x\left(\frac{3}{2}\right)^{x-2}$. What is $f(2)$?

The function f is multiplication of two function $g(x) = 2x$, and $h(x) = \left(\frac{3}{2}\right)^{x-2}$.

Now, to evaluate $f(2)$, you can put the value of $x = 2$ in the function f:

$$x = 2 \rightarrow f(2) = 2(2)\left(\frac{3}{2}\right)^{(2)-2} = 4\left(\frac{3}{2}\right)^0 = 4(1) = 4.$$

In another way, you can find the value of $g(x) = 2x$ and $h(x) = \left(\frac{3}{2}\right)^{x-2}$ for $x = 2$ and multiply the result. Therefore,

$$g(2) = 2 \times 2 = 4$$

$$h(2) = \left(\frac{3}{2}\right)^{2-2} = \left(\frac{3}{2}\right)^0 = 1$$

Finally, $f(2) = g(2) \times h(2) \rightarrow f(2) = 4 \times 1 = 4$.

📝 Evaluate.

27) $f(x) = 3^x$

Find: $f(-1)$

28) $g(x) = \frac{1}{2}(2)^x$

Find: $g(2)$

29) $f(x) = -3\left(\frac{1}{2}\right)^{x-1}$

Find: $f(0)$

30) $r(x) = 5x(2)^x$

Find: $r(-1)$

31) $h(x) = x - \left(\frac{3}{2}\right)^x + 1$

Find: $h(1)$

32) $f(x) = 2^{2x-3}$

Find: $f(a)$

33) $f(x) = 3x\left(\frac{1}{2}\right)^{2x+2}$

Find: $f(2)$

34) $f(x) = 3x\left(\frac{1}{2}\right)^{2x+2}$

Find: $f(4)$

EffortlessMath.com

Function Inverses

A function inverse, also known as the reciprocal function, is a function that "undoes" the effect of another function. If two functions $f(x)$ and $g(x)$ are inverses of each other, then $f(g(x)) = x$ and $g(f(x)) = x$. This means that applying function f followed by function g will give you the original input, and applying function g followed by function f will also give you the original input. For example, the inverse of the function $f(x) = 2x + 3$ is $g(x) = \frac{x-3}{2}$. Because,

$$f(g(x)) = f\left(\frac{x-3}{2}\right) = 2\left(\frac{x-3}{2}\right) + 3 = x$$

$$g(f(x)) = g(2x + 3) = \frac{(2x+3)-3}{2} = \frac{2x}{2} = x$$

It is important to note that not all functions have inverse functions. For a function to have an inverse, it must be a one-to-one function, which means that for every y in the range of the function, there can be only one x in the domain that corresponds to it.

To find the inverse of a function, you can follow these steps:

Step 1: Replace the function notation with $y = f(x)$.

Step 2: Interchange x and y to get $x = f(y)$.

Step 3: Solve for y to get $y = f^{-1}(x)$.

Step 4: Replace y with the original function notation to get the inverse function $f^{-1}(x)$.

It's also worth noting that not all functions have inverse functions. for instance, functions that are not one-to-one, such as $y = x^2$, do not have an inverse function.

Here is an example of how to find the inverse of a function:

Find the inverse of the function $f(x) = 2x + 3$.

Step 1: Replace the function notation with $y = f(x)$ so $y = 2x + 3$.

Step 2: Interchange x and y to get $x = f(y)$ so $x = 2y + 3$.

Step 3: Solve for y to get $y = f^{-1}(x)$: $x = 2y + 3 \rightarrow x - 3 = 2y \rightarrow y = \frac{x-3}{2}$

EffortlessMath.com

Step 4: Replace y with the original function notation to get the inverse function $f^{-1}(x)$ so $f^{-1}(x) = \frac{x-3}{2}$.

So, the inverse of the function $f(x) = 2x + 3$ is $f^{-1}(x) = \frac{x-3}{2}$.

It's worth noting that the function $f(x) = 2x + 3$ is a one-to-one function, so it has an inverse function.

You can also check if the inverse function is correct by plugging in a value for x and seeing if it gives the original input when plugged back into the original function. For example, $f(f^{-1}(3)) = f\left(\frac{x-3}{2}\right) = \left(2\left(\frac{x-3}{2}\right) + 3\right) = 3$, which is the original input.

Other examples:

Find the inverse of the function: $f(x) = \frac{1}{2}x - 1$.

First, replace $f(x)$ with y: $y = \frac{1}{2}x - 1$.

Then, replace all x's with y and all y's with x: $x = \frac{1}{2}y - 1$.

Now, solve for y: $x = \frac{1}{2}y - 1 \to x + 1 = \frac{1}{2}y \to 2(x+1) = y \to y = 2x + 2$.

Finally, replace y with $f^{-1}(x)$: $f^{-1}(x) = 2x + 2$.

Find the inverse of the function: $h(x) = \sqrt{x} + 6$.

Replace the function notation with $y = h(x)$: $h(x) = \sqrt{x} + 6 \to y = \sqrt{x} + 6$.

Next, replace all x's with y and all y's with x;

$x = \sqrt{y} + 6 \to x - 6 = \sqrt{y} \to (x-6)^2 = (\sqrt{y})^2 \to x^2 - 12x + 36 = y$.

Then: $h^{-1} = x^2 - 12x + 36$.

✎ Find the inverse of the following functions.

35) $f(x) = -\frac{1}{x} - 9$

36) $g(x) = \sqrt{x} - 2$

37) $h(x) = -\frac{5}{x+3}$

38) $f(x) = 6x + 6$

Domain and Range of Relations

A relation is a set of ordered pairs that shows the connection or correspondence between elements of two sets. The ordered pair, typically represented as (x, y), contains an input or "x" value and an output or "y" value. Essentially, a relation is a rule that links a value or element from one set to a value or element from another set.

In other words, a relation is defined as a set or the desired set(s) connection. An ordered pair, commonly named as a point and a relation is a set of these ordered pairs. In fact, inputs and outputs values in a relationship is are shown in ordered pairs. Actually, a relation is a kind of rule that connects a component or value from one set to a component or value from the other set.

The domain of a relation is the set of all input (x) values for which the relation is defined. The range of a relation is the set of all output (y) values that can be achieved by the relation.

To find the domain and range of a relation, you can follow these steps:

- Step 1: Identify the set of all x—values in the relation. These values make up the domain of the relation.
- Step 2: Identify the set of all y—values in the relation. These values make up the range of the relation.
- Step 3: Look for any x—values that are not allowed in the relation. These values are excluded from the domain.
- Step 4: Look for any y—values that are not possible in the relation. These values are excluded from the range.

Let's review an example:

Find the domain and range of the relation $y = 2x + 3$.

- Step 1: Identify the set of all x—values in the relation. The domain is all real numbers.

Relations and Functions

Step 2: Identify the set of all y-values in the relation. The range is all real numbers.

Step 3: Look for any x-values that are not allowed in the relation. There are no x-values that are not allowed in the relation.

Step 4: Look for any y-values that are not possible in the relation. There are no y-values that are not possible in the relation.

So, the domain of the relation $y = 2x + 3$ is all real numbers and the range is all real numbers.

Another example:

What is the domain and range of the following relation?
$$\{(7,2), (-3,4), (4,-1), (5,3), (8,5)\}$$

The domain contains x-values of a relation and the range includes y-values of a relationship. So, Domain $= \{7, -3, 4, 5, 8\}$ and Range $= \{2, 4, -1, 3, 5\}$.

Find the domain and range of the following relation:
$$R = \{(4n + 1, n - 2) : n \in \{-1, -2, 0, 2\}\}$$

In the question we have given $n = \{-1, -2, 0, 2\}$. So, put these values in two equations $4n + 1$ and $n - 2$ of the relation R to find the domain and range of the relation:

$$n = -1 \rightarrow \begin{matrix} 4(-1) + 1 = -3 \\ (-1) - 2 = -3 \end{matrix}, \quad \text{and} \quad n = -2 \rightarrow \begin{matrix} 4(-2) + 1 = -7 \\ (-2) - 2 = -4 \end{matrix}$$

$$n = 0 \rightarrow \begin{matrix} 4(0) + 1 = 1 \\ (0) - 2 = -2 \end{matrix}, \quad \text{and} \quad n = 2 \rightarrow \begin{matrix} 4(2) + 1 = 9 \\ (2) - 2 = 0 \end{matrix}$$

Therefore, the domain of the relation R is the set $\{-3, -7, 1, 9\}$ and the range of R is $\{-3, -4, -2, 0\}$.

✏ Write the domain and the range of the following.

39) $\{(1, -1), (2, -4), (0, 5), (-1, 6)\}$

40) $\{(10, -5), (-16, -8), (-4, 19), (16, 7), (6, -14)\}$

41) $\{(4, 7), (-15, 6), (-20, 9), (13, 8), (7, 5)\}$

Interval Notation

Interval notation is a way of expressing a set of numbers in a concise and precise manner. It is commonly used to express the domain and range of a function.

For example, if the domain of a function is all real numbers greater than or equal to 3 and less than or equal to 5, the interval notation for the domain would be $[3,5]$.

The notation for the domain is $[a, b]$, where a and b are the lower and upper bounds of the domain, respectively.

Similarly, if the range of a function is all real numbers greater than -3 and less than 4, the interval notation for the range would be $(-3,4)$.

The notation for the range is (a, b), where a and b are the lower and upper bounds of the range, respectively.

It's important to note that when the interval notation is used to express a set of numbers, the square brackets "[ ]" indicate that the set includes the endpoint values, while the round parentheses "( )" indicate that the set does not include the endpoint values.

For example, the interval notation $[2,5]$ means that the set of numbers includes 2 and 5, while the interval notation $(2,5)$ means that the set of numbers does not include 2 and 5.

Here are some more examples of interval notation for the domain and range of a function:

- If the domain of a function is all real numbers greater than -2 and less than or equal to 3, the interval notation for the domain would be $(-2,3]$.
- If the range of a function is all real numbers greater than or equal to 0 and less than 4, the interval notation for the range would be $[0,4)$.
- If the domain of a function is all real numbers less than or equal to -1, the interval notation for the domain would be $(-\infty, -1]$.

o If the range of a function is all real numbers greater than 2, the interval notation for the range would be $(2, +\infty)$.

It's also possible to use interval notation to express multiple intervals.

To express multiple intervals using interval notation, we can use the union symbol "∪" to combine the different intervals.

For example, consider the following function:
$$f(x) = x^2 - 4x + 2$$
The domain of this function is all real numbers, but the range is restricted to values greater than or equal to -2. The interval notation for the domain would be $(-\infty, +\infty)$ and the interval notation for the range would be $[-2, +\infty)$.

Another example:

Write the set $\{x|-3 < x \leq 1\}$ in the form of an interval notation.

To write this interval in interval notation, use the half-open and half-closed as $(-3, 1]$.

One more example:

Write as an interval notation.

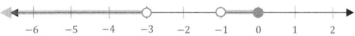

We use the union operator to represented this set in the form of an interval notation: $(-\infty, -3) \cup (-1, 0]$.

✍ Write the following in the form of interval notation.

42) The set of all real numbers less than or equal to 1 and greater than -1.

43) The set of all real numbers greater than or equal to 3.

44) The set $\{x \in \mathbb{R} | 1 < x < 4, -2 \leq x \leq -1\}$.

✍ Write the domain and the following as an interval.

45) $f(x) = (x+1)^2$ 46) $g(x) = 3x - 1$

Rate of Change and Slope

The rate of change, also known as the slope, is a ratio used to compare the change in y-values to the change in x-values. The y-values, also known as dependent variables, are dependent on the x-values, or independent variables. When the rate of change is constant and linear, the slope of a line can be represented as a number. The slope can be positive, negative, zero, or undefined.

The direction of a line's slope can also be described by its slope. A line with a positive slope rises from left to right, while a line with a negative slope falls from left to right. A line with a slope of zero is a horizontal line, indicating no change in the y-value. A vertical line, on the other hand, represents a relationship that is not a function, and the slope is undefined as there are many y-values for one x-value. It's because there are many y-values for one x-value.

The variable m is used to represent the slope of a line. It is calculated as the ratio of the change in y-values to the change in x-values. If the coordinates of two points on a line are (x_1, y_1) and (x_2, y_2), then the slope m is calculated as

$$m = \frac{y_2 - y_1}{x_2 - x_1}$$

For example, consider the line represented by the equation $y = 2x + 3$.

In this equation, the slope m is 2, which means for every unit increase in x, the y-value increases by 2 units. This slope is positive, indicating that the line rises from left to right.

You can also find the slope of a line by finding the slope between two points on the line. For example, if we have two points on this line, $(2,7)$ and $(5,13)$. The coordinates of these two points are $(x_1, y_1) = (2,7)$ and $(x_2, y_2) = (5,13)$. The slope between these two points is $m = \frac{y_2 - y_1}{x_2 - x_1} = \frac{13 - 7}{5 - 2} = \frac{6}{3} = 2$, which is the same as the slope of the line represented by the equation $y = 2x + 3$.

 Let's review an example:

Relations and Functions

The following table shows the number of cars sold by a company in different years. Find the rate of change in car sales for each time interval. Determine which time interval has the greatest rate.

Year	2005	2010	2015	2020	2022
Number of Sold Cars	35	45	47	67	85

First, find the dependent and independent variables: years are independent variables and the number of sold cars are dependent variables. Now, find the rate of changes:

2005 to 2010: $\frac{change\ in\ the\ number\ of\ sold\ cars}{change\ in\ years} = \frac{45-35}{2010-2005} = 2$.

2010 to 2015: $\frac{change\ in\ the\ number\ of\ sold\ cars}{change\ in\ years} = \frac{47-45}{2015-2010} = 0.4$.

2015 to 2020: $\frac{change\ in\ the\ number\ of\ sold\ cars}{change\ in\ years} = \frac{67-47}{2020-2015} = 4$.

2020 to 2022: $\frac{change\ in\ the\ number\ of\ sold\ cars}{change\ in\ years} = \frac{85-67}{2022-2020} = 9$.

Car sales have been at their greatest rate from 2022 to 2020. The slope of the line is positive because as time increases, the amount of car sales also increases.

✍ Solve.

47) Average food preparation time in a restaurant was tracked daily as part of an efficiency improvement program. According to the table, what was the rate of change between Tuesday and Wednesday?

Day	Time
Tuesday	45
Wednesday	49
Thursday	32
Friday	15
Saturday	25

Complete a Function Table from an Equation

A function table is a useful tool to demonstrate the relationship between input and output values in a mathematical equation. To complete a function table:

Step 1: Create a table with columns for the input variable (*x*) and the output variable (*y*).

Step 2: Fill in the input values in the column *x*.

Step 3: Substitute the input values into the equation.

Step 4: Evaluate the equation to find the corresponding output values and fill them in the column *y*.

Here is an example of how to complete a function table from an equation:

Complete the function table for the equation $y = 2x + 3$.

$f(x) = 2x + 3$	
x	$f(x)$
0	
1	
2	

Write down the equation $y = 2x + 3$.

Step 1: Write down a table with two columns, one for *x* and one for *y*.

Step 2: Insert a set of values for *x* in the *x* column of the table.

Step 3: Substitute each *x* value into the equation and solve for *y*.

Step 4: Insert the corresponding *y* values into the column *y* of the table.

$f(x) = 2x + 3$	
x	$f(x)$
0	3
1	5
2	7

Check if the input *x* value correctly corresponds to the output *y* value.

By following these steps, you can complete a function table from an equation and check if the relationship between *x* and *y* is correct.

Let's review an example:

Complete the table.

$f(x) = x^2 + 1$	
x	$f(x)$
-2	
0	
1	

Look at the function table. Clearly, the first value of the input in the table is -2. Evaluate $f(x) = x^2 + 1$ for $x = -2$. $f(-2) = (-2)^2 + 1 = 4 + 1 = 5$.

When $x = -2$, then $f(-2) = 5$. Complete the first row of the table.

$f(x) = x^2 + 1$	
x	$f(x)$
-2	5
0	
1	

Similarly, evaluate $f(x) = x^2 + 1$ for $x = 0$ and $x = 1$, respectively. So,

$f(0) = (0)^2 + 1 = 0 + 1 = 1$,

$f(1) = (1)^2 + 1 = 1 + 1 = 2$.

Enter the obtained values in the table.

$f(x) = x^2 + 1$	
x	$f(x)$
-2	5
0	1
1	2

✎ Complete the tables.

48)

$f(x) = 3x - 2$	
x	$f(x)$
-3	
0	
2	

49)

$f(x) = 2x$	
x	$f(x)$
1	
2	
3	

Chapter 8: Answers

1) -3
2) 10
3) -10
4) 3
5) 37
6) -1
7) $8n^2 - 8n$
8) $-108a^3 - 10$
9) 13
10) 0
11) 16
12) 29
13) $3x^2 - 12x + 22$
14) $-6a^2 - 33$
15) -30
16) $\frac{1}{2}$
17) $\frac{14}{3}$
18) 49
19) -160
20) -4
21) 10
22) -5
23) 16
24) 6
25) 8
26) 19
27) $\frac{1}{3}$
28) 2
29) -6
30) $-\frac{5}{2}$
31) $\frac{1}{2}$
32) 2^{2a-3}
33) $\frac{3}{32}$
34) $\frac{3}{256}$
35) $-\frac{1}{x+9}$
36) $x^2 + 4x + 4$
37) $-\frac{5}{x} - 3$
38) $\frac{x-6}{6}$
39) $D = (1, 2, 0, -1)$, $R = (-1, -4, 5, 6)$
40) $D = (10, -16, -4, 16, 6)$, $R = (-5, -8, 19, 7, -14)$
41) $D = (4, -15, -20, 13, 7)$, $R = (7, 6, 9, 8, 5)$
42) $(-1, 1]$

Relations and Functions

43) $[3, +\infty)$

44) $[-2, -1] \cup (1, 4)$

45) $(-\infty, +\infty)$

46) $(-\infty, +\infty)$

47) 4

48)

$f(x) = 3x - 2$	
x	$f(x)$
−3	−11
0	−2
2	4

49)

$f(x) = 2x$	
x	$f(x)$
1	2
2	4
3	6

CHAPTER 9
Radical Expressions

Math topics that you'll learn in this chapter:

- ☑ Simplifying Radical Expressions
- ☑ Adding and Subtracting Radical Expressions
- ☑ Multiplying Radical Expressions
- ☑ Rationalizing Radical Expressions
- ☑ Radical Equations
- ☑ Domain and Range of Radical Functions
- ☑ Simplify Radicals with Fractions

Simplifying Radical Expressions

Here are some steps to simplify radical expressions:

Step 1: Factor the radicand (the number or expression inside the radical symbol) as much as possible.

Step 2: Identify any perfect square factors (such as 4, 9, 16, etc.) in the radicand. The square root of a perfect square is the number itself.

Step 3: Use the product property of square roots, which states that the square root of a product is equal to the product of the square roots.

$$\sqrt[n]{xy} = \sqrt[n]{x} \times \sqrt[n]{y}$$

Step 4: Use the quotient property of square roots, which states that the square root of a quotient is equal to the quotient of the square roots.

$$\sqrt[n]{\frac{x}{y}} = \frac{\sqrt[n]{x}}{\sqrt[n]{y}}$$

Step 5: Simplify any radicals that have the same radicand by adding or subtracting them.

Step 6: Simplify any radicals that have the same index by multiplying or dividing them.

Note: When simplifying radical expressions, pay attention to converting radical expressions to exponential expressions, and vice versa:

$$\sqrt[n]{x^a} = x^{\frac{a}{n}}$$

By following these steps, you can simplify radical expressions and make them easier to work with and understand.

Radical Expressions

Let's review an example:

Find the square root of $\sqrt{144x^2}$.

Find the factor of the expression $144x^2$: $144 = 12 \times 12$ and $x^2 = x \times x$, now use radical rule: $\sqrt[n]{a^n} = a$.

Then: $\sqrt{12^2} = 12$ and $\sqrt{x^2} = x$.

Finally: $\sqrt{144x^2} = \sqrt{12^2} \times \sqrt{x^2} = 12 \times x = 12x$.

Another example:

Write this radical in exponential form: $\sqrt[3]{x^4}$.

To write a radical in exponential form, use this rule: $\sqrt[n]{x^a} = x^{\frac{a}{n}}$.

Then: $\sqrt[3]{x^4} = x^{\frac{4}{3}}$.

One more example:

Simplify: $\sqrt{8x^3}$.

First, factor the expression $8x^3$: $8x^3 = 2^3 \times x \times x \times x$.

We need to find perfect squares: $8x^3 = 2^2 \times 2 \times x^2 \times x = 2^2 \times x^2 \times 2x$.

Then: $\sqrt{8x^3} = \sqrt{2^2 \times x^2} \times \sqrt{2x}$.

Now, use the radical rule: $\sqrt[n]{a^n} = a$.

Then: $\sqrt{2^2 \times x^2} \times \sqrt{(2x)} = 2x \times \sqrt{2x} = 2x\sqrt{2x}$.

✎ Evaluate.

1) $\sqrt{49} = $ _____

2) $\sqrt{4} \times \sqrt{81} = $ _____

3) $\sqrt{16} \times \sqrt{4x^2} = $ _____

4) $\sqrt{289} = $ _____

5) $\sqrt{25b^4} = $ _____

6) $\sqrt{9} \times \sqrt{x^2} = $ _____

Adding and Subtracting Radical Expressions

When adding or subtracting radical expressions, it's important to remember the following rules:

1. The radicand (the expression inside the radical symbol) must be the same to combine them.

2. To add or subtract radical expressions, simply add or subtract the numerical coefficients in front of each radical. For example:

   $\sqrt{9x^2} + 2\sqrt{9x^2} = (1+2)\sqrt{9x^2} = 3\sqrt{9x^2} = 9x.$

3. When the radicand is not the same it is not possible to combine them, leaving the radical expressions as they are. For example: $\sqrt{5x^2} + 2\sqrt{3x}$, cannot be combined.

Here are a few more examples of adding and subtracting radical expressions:

Simplify: $\sqrt{48} + 2\sqrt{48}$.

In this example, the radicand is the same $\sqrt{48}$. So, we can add the numerical coefficients in front of the radical $(1 + 2)$ to get 3 and combine them.

Therefore, we get:

$\sqrt{48} + 2\sqrt{48} = (1+2)\sqrt{48} = 3\sqrt{48}.$

Note that the number of 48 can be written as the following product:

$48 = 4^2 \times 3.$

Next, we have: $\sqrt{48} = \sqrt{4^2 \times 3} = 4\sqrt{3}$. Therefore, substitute:

$\sqrt{48} + 2\sqrt{48} = 3\sqrt{48} = 3 \times 4\sqrt{3} = 12\sqrt{3}.$

Another example:

Simplify: $2\sqrt[3]{3x^2} - 3\sqrt[3]{3x^2}$.

Radical Expressions

In this example, the radicand is the same $3x^2$. So, we can add the numerical coefficients in front of the radical $(2-3)$ to get -1 and combine them.

Therefore, we have: $2\sqrt[3]{3x^2} - 3\sqrt[3]{3x^2} = (2-3)\sqrt[3]{3x^2} = -\sqrt[3]{3x^2}$.

Some other examples:

Simplify: $\sqrt{27x} + 2\sqrt{3x}$.

The two radical parts are not the same. Then, rewrite the expression as:

$\sqrt{27x} + 2\sqrt{3x} = \sqrt{(9 \times 3)x} + 2\sqrt{3x} = 3\sqrt{3x} + 2\sqrt{3x}$

Since we have the same radical parts, then we can add these two radicals.

Add like terms: $3\sqrt{3x} + 2\sqrt{3x} = (3+2)\sqrt{3x} = 5\sqrt{3x}$.

Simplify: $2\sqrt{8} - 2\sqrt{2}$.

The two radical parts are not the same.

First, we need to simplify the $2\sqrt{8}$. Then: $2\sqrt{8} = 2\sqrt{4 \times 2} = 2(\sqrt{4})(\sqrt{2}) = 4\sqrt{2}$.

Now, combine like terms: $2\sqrt{8} - 2\sqrt{2} = 4\sqrt{2} - 2\sqrt{2} = 2\sqrt{2}$.

Simplify: $8\sqrt{27} + 5\sqrt{3}$.

The two radical parts are not the same.

First, we need to simplify the $8\sqrt{27}$. Then: $8\sqrt{27} = 8\sqrt{9 \times 3} = 8(\sqrt{9})(\sqrt{3}) = 24\sqrt{3}$.

Now, add: $8\sqrt{27} + 5\sqrt{3} = 24\sqrt{3} + 5\sqrt{3} = 29\sqrt{3}$.

✍ Simplify.

7) $\sqrt{6} + 6\sqrt{6} =$

8) $9\sqrt{8} - 6\sqrt{2} =$

9) $-\sqrt{7} - 5\sqrt{7} =$

10) $10\sqrt{2} + 3\sqrt{18} =$

11) $\sqrt{12} - 6\sqrt{3} =$

12) $-2\sqrt{x} + 6\sqrt{x} =$

Multiplying Radical Expressions

When multiplying radical expressions, it's important to remember the following rules:

1. To multiply radical expressions, multiply the numerical coefficients in front of the radical and multiply the radicands (the expressions inside the radical symbol), separately.
2. When there are radicals with the same radicand, we can simplify the radical by adding the indices (the number written just outside or on the radical symbol).

Here are a few examples of multiplying radical expressions:

Evaluate: $2\sqrt{3x} \times 3\sqrt{5x}$.

In this example, we multiply the numerical coefficients in front of the radical (2×3) and multiply the radicands $(3x \times 5x)$, separately to get:

$2\sqrt{3x} \times 3\sqrt{5x} = 6\sqrt{15x^2} = 6x\sqrt{15}$.

Evaluate. $2\sqrt{5} \times \sqrt{3}$

Multiply the numbers outside of the radicals (2×1) and the radical parts (5×3). Then: $2\sqrt{5} \times \sqrt{3} = (2 \times 1) \times (\sqrt{5} \times \sqrt{3}) = 2\sqrt{15}$.

Simplify: $3\sqrt{2x} \times 4\sqrt{2x}$.

For this purpose, we multiply the numerical coefficients in front of the radical (3×4) and multiply the radicands $(2x \times 2x)$, separately to get:

$3\sqrt{2x} \times 4\sqrt{2x} = (3 \times 4)\sqrt{2x \times 2x} = 12\sqrt{4x^2} = 24x$.

Simplify: $3x\sqrt{3} \times 4\sqrt{x}$.

Multiply the numbers outside of the radicals $(3x \times 4)$ and the radical parts $(3 \times x)$. Then, substitute and simplify:

$3x\sqrt{3} \times 4\sqrt{x} = (3x \times 4) \times (\sqrt{3} \times \sqrt{x}) = (12x)(\sqrt{3 \times x}) = 12x\sqrt{3x}$.

Evaluate the product of the radical expression: $(2\sqrt{5})^3 =$.

Raise the coefficient of the radical to power 3 and multiply it with the radicand raised to power 3: $(2\sqrt{5})^3 = (2)^3 \times (\sqrt{5})^3$. So, we have: $(2 \times 2 \times 2)\sqrt{5 \times 5 \times 5} = 8\sqrt{125}$

Next, simplify: $8\sqrt{125} = 8 \times 5\sqrt{5} = 40\sqrt{5}$.

Therefore, the product of this expression is $(2\sqrt{5})^3 = 40\sqrt{5}$.

Evaluate: $(2\sqrt{4x})^2$.

We raise coefficient of the radical to the power 2 and multiply it with the radicand raised to power 2. So, we get:

$(2\sqrt{4x})^2 = (2)^2 \times (\sqrt{4x})^2 = 4\sqrt{(4x)^2} = 4\sqrt{16x^2} = 4 \times 4x = 16x$.

Multiply $6a\sqrt{7b}$ to $3\sqrt{2b}$ and simplify.

Multiply the numbers outside of the radicals ($6a \times 3$) and the radical parts ($7b \times 2b$). Then: $6a\sqrt{7b} \times 3\sqrt{2b} = (6a \times 3) \times \sqrt{7b \times 2b} = 18a\sqrt{14b^2}$.

Next, simplify: $18a\sqrt{14b^2} = 18a \times \sqrt{14} \times \sqrt{b^2} = 18ab\sqrt{14}$.

Simplify: $9\sqrt{9x} \times 5\sqrt{4x}$.

Multiply the numbers outside of the radicals (9 and 5), and the radical parts ($9x$ and $4x$). Then, substitute and simplify:

$9\sqrt{9x} \times 5\sqrt{4x} = (9 \times 5) \times (\sqrt{9x \times 4x}) = (45)(\sqrt{36x^2}) = 45\sqrt{36x^2}$.

$\sqrt{36x^2} = 6x$, then: $45\sqrt{36x^2} = 45 \times 6x = 270x$.

✎ Evaluate.

13) $\sqrt{4} \times 2\sqrt{9} =$

15) $-6\sqrt{4} \times 3\sqrt{4} =$

14) $\sqrt{5y} \times 3\sqrt{20y} =$

16) $-9\sqrt{3b^2} \times (-\sqrt{6}) =$

bit.ly/3ri1RqN

Rationalizing Radical Expressions

Rationalizing radical expressions is the process of eliminating a radical in the denominator of a fraction. In other words, it's a way to convert an expression with a radical in the denominator into an equivalent expression that does not have a radical in the denominator. This is done by multiplying the numerator and denominator by some expression that will eliminate the radical in the denominator.

To rationalize a denominator that contains a radical, we can use the conjugate of the denominator. The conjugate of a binomial is the binomial with the same signs between the two terms but with the opposite sign on the radical. For example, the conjugate of $\sqrt{2} - 2$ is $\sqrt{2} + 2$. The product of the original binomial and its conjugate is a perfect square.

For example, simplify: $\frac{x}{\sqrt{2}-2}$.

To rationalize the denominator $\sqrt{2} - 2$, you can multiply the numerator and denominator by the conjugate of the denominator, which is $\sqrt{2} + 2$. The result is:

$$\frac{x}{\sqrt{2}-2} \times \frac{\sqrt{2}+2}{\sqrt{2}+2} = \frac{x(\sqrt{2}+2)}{(\sqrt{2}-2)(\sqrt{2}+2)} = \frac{x\sqrt{2}+2x}{2-4} = \frac{2x+x\sqrt{2}}{-2} = -\frac{2x+x\sqrt{2}}{2}.$$

In the same way, you can rationalize any denominator containing radicals, it's just about multiplying the numerator and denominator by the conjugate of the denominator to eliminate the radical in the denominator.

- Radical expressions cannot be in the denominator. (Number in the bottom)
- To get rid of the radical in the denominator, multiply both the numerator and denominator by the radical in the denominator.

Radical Expressions

- If there is a radical and another integer in the denominator, multiply both the numerator and denominator by the conjugate of the denominator.
- The conjugate of $(a + b)$ is $(a - b)$ and vice versa.
- Rationalizing radical expressions is the process of eliminating a radical in the denominator of a fraction.

Here are a few examples of how to rationalize radical expressions:

Simplify: $\frac{9}{\sqrt{3}}$.

Multiply both the numerator and denominator by $\sqrt{3}$. Then: $\frac{9}{\sqrt{3}} \times \frac{\sqrt{3}}{\sqrt{3}} = \frac{9\sqrt{3}}{\sqrt{9}} = \frac{9\sqrt{3}}{3}$.

Now, simplify: $\frac{9\sqrt{3}}{3} = 3\sqrt{3}$.

Simplify $\frac{5}{\sqrt{6}-4}$.

Multiply by the conjugate: $\frac{\sqrt{6}+4}{\sqrt{6}+4} \to \frac{5}{\sqrt{6}-4} \times \frac{\sqrt{6}+4}{\sqrt{6}+4}$. That is, $(\sqrt{6} - 4)(\sqrt{6} + 4) = -10$.

Then: $\frac{5}{\sqrt{6}-4} \times \frac{\sqrt{6}+4}{\sqrt{6}+4} = \frac{5(\sqrt{6}+4)}{-10}$.

Use the fraction rule: $\frac{a}{-b} = -\frac{a}{b} \to \frac{5(\sqrt{6}+4)}{-10} = -\frac{5(\sqrt{6}+4)}{10} = -\frac{1}{2}(\sqrt{6} + 4)$.

Simplify $\frac{2}{\sqrt{3}-1}$.

Multiply by the conjugate: $\frac{\sqrt{3}+1}{\sqrt{3}+1}$. So, $\frac{2}{\sqrt{3}-1} \times \frac{\sqrt{3}+1}{\sqrt{3}+1} = \frac{2(\sqrt{3}+1)}{2} = (\sqrt{3} + 1)$.

✎ Simplify.

17) $\frac{1}{\sqrt{2}} =$

18) $\frac{1+\sqrt{5}}{1-\sqrt{3}} =$

19) $\frac{2+\sqrt{6}}{\sqrt{2}-\sqrt{5}} =$

20) $\frac{4}{\sqrt{8}} =$

21) $\frac{\sqrt{7}}{\sqrt{6}-\sqrt{3}} =$

22) $\frac{\sqrt{8a}}{\sqrt{a^5}} =$

Radical Equations

A radical equation is an equation in which the independent variable is under the radical. For example:

$$\sqrt{x-1} = 2, \sqrt[3]{x} - \sqrt{2x+3} = 0, \text{ and } \ldots$$

To solve a radical equation, follow the steps below:

Step 1: Isolate the radical on one side of the equation. If more than one radical expression involves the variable, then isolate one of them.

Step 2: Square both sides of the equation to remove the radical.

Step 3: Solve the equation for the variable.

Step 4: Plug the answer (answers) into the original equation to avoid extraneous values.

Let's review some examples:

Solve: $\sqrt{x} - 5 = 15$.

Add 5 to both sides:

$\sqrt{x} - 5 = 15 \rightarrow (\sqrt{x} - 5) + 5 = 15 + 5 \rightarrow \sqrt{x} = 20$.

Square both sides: $(\sqrt{x})^2 = 20^2 \rightarrow x = 400$.

Plugin the value of 400 for x in the original equation and check the answer:

$x = 400 \rightarrow \sqrt{x} - 5 = \sqrt{400} - 5 = 20 - 5 = 15$.

So, the value of 400 for x is correct.

Another example:

What is the value of x in this equation? $2\sqrt{x+1} = 4$

Divide both sides by 2. Then:

$2\sqrt{x+1} = 4 \rightarrow \frac{2\sqrt{x+1}}{2} = \frac{4}{2} \rightarrow \sqrt{x+1} = 2$.

Radical Expressions

Square both sides: $(\sqrt{x+1})^2 = 2^2$.

Then, $x + 1 = 4 \to x = 3$.

Substitute x by 3 in the original equation and check the answer:

$x = 3 \to 2\sqrt{x+1} = 2\sqrt{3+1} = 2\sqrt{4} = 2(2) = 4$.

So, the value of 3 for x is correct.

One more example:

Solve: $\sqrt{x+3} - x = 1$.

First, isolate the radical expression on one side of the equation:

$\sqrt{x+3} - x = 1 \to \sqrt{x+3} = x + 1$.

Next, square both sides and simplify:

$(\sqrt{x+3})^2 = (x+1)^2 \to x + 3 = x^2 + 2x + 1 \to x^2 + x - 2 = 0$

Now, Solve for x. Find the factors of the quadratic equation:

$x^2 + x - 2 = 0 \to (x-1)(x+2) = 0$

Then, the value of x is 1 or -2.

Check the answers by substituting in the original equation to confirm:

$x = 1 \to \sqrt{x+3} - x = \sqrt{1+3} - 1 = \sqrt{4} - 1 = 2 - 1 = 1$

$x = -2 \to \sqrt{x+3} - x = \sqrt{(-2)+3} - (-2) = \sqrt{1} + 2 = 1 + 3 = 4 \neq 1$.

Therefore, only the value $x = 1$ is the correct answer.

✒ Solve for x in each equation.

23) $2\sqrt{2x-4} = 8$

24) $9 = \sqrt{4x-1}$

25) $\sqrt{x} + 6 = 11$

26) $\sqrt{5x} = \sqrt{x+3}$

27) $\sqrt[3]{3x+2} = 2$

28) $\sqrt{1-2x} - 2\sqrt{x} = 0$

Domain and Range of Radical Functions

To find the domain of the radical function, find all possible values of the variable inside the radical.

Remember that having a negative number under the square root symbol is not possible. (For cubic roots, we can have negative numbers.)

To find the range, plugin the minimum and maximum values of the variable inside radical.

In order to find the domain and range of a radical function in the form,

$$f(x) = c\sqrt[m]{g(x)} + k$$

Where m is a positive integer and $g(x)$ is a function. $c \neq 0$ and k are constant number. Go through the following steps:

Step 1: Determine whether the radical index is even or odd.

Step 2: To find the domain of the radical function, we evaluate the possible values for the input x as follows:

If m is even, then the domain can be identified by finding the values of x satisfying $g(x) \geq 0$.

If m is odd, then the domain can be identified by finding the values of x for which $g(x)$ is defined.

Step 3: The range of the function f is the intersection of the set of all possible values for g with:

For m is even. Since $\sqrt[m]{g(x)} \geq 0$, then:

$$c > 0: c\sqrt[m]{g(x)} \geq 0 \to c\sqrt[m]{g(x)} + k \geq k \to f(x) \geq k$$
$$c < 0: c\sqrt[m]{g(x)} \leq 0 \to c\sqrt[m]{g(x)} + k \leq k \to f(x) \leq k$$

For m is odd. The expression $\sqrt[m]{g(x)}$ can be any real number for all the possible values of x, then $f(x) \in \mathbb{R}$.

Step 4: Write the final answer in inequality notation or interval notation.

Let's review an example:

Find the domain and range of the radical function. $y = \sqrt{x-3}$

Since the radical index is even (2).

Then the domain is the values of x such that: $x - 3 \geq 0 \rightarrow x \geq 3$.

It means that the domain is the interval $[3, +\infty)$.

For the range of the function, we know that $\sqrt{x-3} \geq 0$. So, $y \geq 0$. Therefore, the range is the interval $[0, +\infty)$.

Another example:

Find the domain and range of the radical function. $y = -6\sqrt{4x+8} + 5$

For the domain, find non-negative values for radicals: $4x + 8 \geq 0$.

Therefore, $4x + 8 \geq 0 \rightarrow 4x \geq -8 \rightarrow x \geq -2$.

The domain of the function $y = 6\sqrt{4x+8} + 5$ is all values of real numbers that are satisfied in the inequality $x \geq -2$. Equivalent to the interval $[-2, +\infty)$.

For range, we know that $\sqrt{4x+8} \geq 0$. Then,

$-6\sqrt{4x+8} \leq 0 \rightarrow -6\sqrt{4x+8} + 5 \leq 5 \rightarrow y \leq 5$.

The range of the function $y = -6\sqrt{4x+8} + 5$ is $y \leq 5$. Equivalent to the interval $(-\infty, 5]$.

✎ Identify the domain and range of each function.

29) $y = \sqrt{x+1}$

30) $y = \sqrt{x-2} + 6$

31) $y = \sqrt{x} - 1$

32) $y = \sqrt{x-4}$

Simplify Radicals with Fractions

To simplify radicals with fractions:

- Rewrite the numerator and denominator of the fraction as the product of the prime factorizations. (Write in the product of prime numbers.)

    For example, the number 19 is a prime number (factors of 19 are 1 and 19). The factors of the number 21 are 1, 3, 7 and 21.

    $$21 = 3 \times 7$$

- Apply the multiplication and division properties of radical expressions.

    For $x, y \geq 0$:

    $$\sqrt[n]{x \times y} = \sqrt[n]{x} \times \sqrt[n]{y}$$

    And $x \geq 0, y > 0$:

    $$\sqrt[n]{\frac{x}{y}} = \frac{\sqrt[n]{x}}{\sqrt[n]{y}}$$

- Group the factors that form a perfect square, perfect cube, etc.

    For example, the factors of the number 360 can be grouped as follows:

    $$360 = 2 \times 2 \times 2 \times 5 \times 3 \times 3 = 2^3 \times 5 \times 3^2$$

- Simplify.

Let's review some examples to further explanation:

Simplify: $\sqrt{\frac{9}{25}}$.

To simplify the radical fraction, rewrite the numerator and denominator as the product of the prime factorizations.

　　The product of the factors of 9 is 3×3.

　　The product of the factors of 25 is 5×5.

Radical Expressions

So, $\sqrt{\frac{9}{25}} = \sqrt{\frac{3 \times 3}{5 \times 5}}$.

Since the index of the given radical is 2. You can take one term out of radical for every two same terms multiplied inside the radical sign (perfect square): $9 = 3^2$, and $25 = 5^2$. Then: $\sqrt{\frac{9}{25}} = \sqrt{\frac{3^2}{5^2}} = \frac{3}{5}$.

Simplify the following radical fraction: $\sqrt{\frac{44}{16}}$.

Rewrite the numerator and denominator of the fraction as follow: $\sqrt{\frac{44}{16}} = \sqrt{\frac{2 \times 2 \times 11}{2 \times 2 \times 2 \times 2}}$

Consider the index of the given radical, take out one term of radical for every term that is repeated in an even number inside the radical sign. So,

$\sqrt{\frac{2 \times 2 \times 11}{2 \times 2 \times 2 \times 2}} = \sqrt{\frac{2^2 \times 11}{2^4}} = \frac{2\sqrt{11}}{2^2}$.

Simplify, $\frac{2\sqrt{11}}{2^2} = \frac{\sqrt{11}}{2}$.

Write the expression in the simplest radical form: $\sqrt{\frac{242}{45}}$.

Rewrite this radical fraction as the product of the prime factorizations:

$\sqrt{\frac{242}{45}} = \sqrt{\frac{2 \times 11 \times 11}{3 \times 3 \times 5}}$.

Now, take out the terms that are perfect squares, so, $\sqrt{\frac{2 \times 11 \times 11}{3 \times 3 \times 5}} = \sqrt{\frac{2 \times 11^2}{3^2 \times 5}} = \frac{11}{3}\sqrt{\frac{2}{5}}$.

✎ Solve.

33) $\sqrt{\frac{625}{36}}$

34) $\sqrt{\frac{1296}{25}}$

35) $\sqrt{\frac{147}{64}}$

36) $\sqrt{\frac{98}{18}}$

Chapter 9: Answers

1) 7
2) 18
3) $8x$
4) 17
5) $5b^2$
6) $3x$
7) $7\sqrt{6}$
8) $12\sqrt{2}$
9) $-6\sqrt{7}$
10) $19\sqrt{2}$
11) $-4\sqrt{3}$
12) $4\sqrt{x}$
13) 12
14) $30y$
15) -72
16) $27b\sqrt{2}$
17) $\frac{\sqrt{2}}{2}$
18) $-\frac{(1+\sqrt{5})(1+\sqrt{3})}{2}$
19) $-\frac{2\sqrt{2}+2\sqrt{5}+2\sqrt{3}+\sqrt{30}}{3}$
20) $\sqrt{2}$
21) $\frac{\sqrt{7}(\sqrt{6}+\sqrt{3})}{3}$
22) $\frac{2\sqrt{2}}{a^2}$
23) $x = 10$
24) $x = 20.5$
25) $x = 25$
26) $x = \frac{3}{4}$
27) $x = 2$
28) $x = \frac{1}{6}$
29) $x \geq -1, y \geq 0$
30) $x \geq 2, y \geq 6$
31) $x \geq 0, y \geq -1$
32) $x \geq 4, y \geq 0$
33) $\frac{25}{6}$
34) $\frac{36}{5}$
35) $\frac{7\sqrt{3}}{8}$
36) $\frac{7}{3}$

Chapter 10: Rational Expressions

Math topics that you'll learn in this chapter:

- ☑ Simplifying Complex Fractions
- ☑ Graphing Rational Functions
- ☑ Adding and Subtracting Rational Expressions
- ☑ Multiplying Rational Expressions
- ☑ Dividing Rational Expressions
- ☑ Evaluate Integers Raised to Rational Exponents

Simplifying Complex Fractions

A complex fraction is a fraction that contains one or more other fractions in either the numerator or denominator. To simplify a complex fraction, the first step is to find a common denominator for all the fractions involved. A common denominator is a number that is a multiple of all the denominators of the fractions. Once you have a common denominator, you can add or subtract the numerators as appropriate, and then simplify the resulting fraction if possible.

For example, consider the complex fraction $\frac{2}{3} + \frac{4}{5}$.

To simplify this fraction, we first need to find a common denominator. In this case, a common denominator would be 15. Since 15 is a multiple of both 3 and 5. Once we have a common denominator, we can rewrite the fractions. So, that they have the same denominator.

In this case, we would have $\frac{2}{3} = \frac{10}{15}$ and $\frac{4}{5} = \frac{12}{15}$. With these equivalent fractions, we can add the numerators: 10 + 12 = 22, and the resulting fraction is $\frac{22}{15}$.

$$\frac{2}{3} + \frac{4}{5} = \frac{10}{15} + \frac{12}{15} = \frac{10 + 12}{15} = \frac{22}{15}$$

It's important to note that simplifying complex fractions doesn't always result in a fraction with a smaller numerator and denominator. The resulting fraction may have the same numerator and denominator or even greater numerator and denominator.

Let's review an example:

Simplify: $\frac{\frac{3}{5}}{\frac{3}{5} - \frac{5}{8}} =$.

To simplify the complex fraction $\frac{\frac{3}{5}}{\frac{3}{5} - \frac{5}{8}}$, we will first simplify each fraction within the complex fraction. First, we simplify the denominator of the complex fraction:

$$\frac{3}{5} - \frac{5}{8} = \frac{3 \times 8}{5 \times 8} - \frac{5 \times 5}{8 \times 5} = \frac{24}{40} - \frac{25}{40} = -\frac{1}{40}.$$

Rational Expressions

Now, substitute this simplified denominator into the original complex fraction:

$$\frac{\frac{3}{5}}{-\frac{1}{40}} = \left(\frac{3}{5}\right) \times \left(-\frac{40}{1}\right) = -\frac{120}{5}.$$

And we can simplify further if we want to, we can divide both the numerator and denominator by a common factor of 5: $-\frac{120}{5} = -24$.

So, the simplified form of the complex fraction is -24.

Simplify: $\frac{\frac{2}{7} \div \frac{1}{4}}{\frac{7}{8} + \frac{1}{4}}$.

First, simplify the numerator: $\frac{2}{7} \div \frac{1}{4} = \frac{8}{7}$.

Then, simplify the denominator: $\frac{7}{8} + \frac{1}{4} = \frac{9}{8}$.

Now, write the complex fraction using the division sign (÷):

$$\frac{\frac{2}{7} \div \frac{1}{4}}{\frac{7}{8} + \frac{1}{4}} = \frac{\frac{8}{7}}{\frac{9}{8}} = \frac{8}{7} \div \frac{9}{8}.$$

Use the dividing fractions rule: (Keep, Change, Flip: Keep the first fraction, Change the division sign to multiplication, Flip the second fraction)

$$\frac{8}{7} \div \frac{9}{8} = \frac{8}{7} \times \frac{8}{9} = \frac{64}{63} = 1\frac{1}{63}.$$

✎ Simplify each expression.

1) $\frac{\frac{2}{5}}{\frac{4}{3} + \frac{6}{5}} =$

2) $\frac{3}{\frac{1}{4} - \frac{3}{5}} =$

3) $\frac{\frac{3}{4} - \frac{2}{9}}{\frac{3}{8} + \frac{5}{9}} =$

4) $\frac{\frac{5}{21} - \frac{2}{3}}{\frac{6}{7} + \frac{4}{5}} =$

5) $\frac{\frac{18}{32}}{6} =$

6) $\frac{\frac{9}{15} + \frac{4}{5}}{\frac{36}{25}} =$

7) $\frac{\frac{8}{3}}{\frac{2}{5}} =$

8) $\frac{\frac{x}{3} + \frac{x}{8}}{\frac{1}{4}} =$

9) $\frac{\frac{x+3}{3}}{\frac{x-2}{2}} =$

10) $\frac{1 + \frac{x}{4}}{x} =$

Graphing Rational Functions

Graphing rational functions, which are functions of the form $f(x) = \frac{P(x)}{Q(x)}$, where $P(x)$ and $Q(x)$ are polynomials, can be done by following these steps:

Step 1: Find the asymptotes: Determine each of the types of asymptotes. There are three types of asymptotes for rational functions:
- Vertical asymptotes: These are the x-values at which the denominator of the function is equal to zero.
- Horizontal asymptotes: These are the y-values that the graph approaches as x approaches positive or negative infinity.
- Slanting asymptote: It is a slanted line with the equation of the form $y = mx + b$.

Step 2: Find the x-intercepts: These are the x-coordinates of the points where the graph crosses the x-axis.

Step 3: Find the y-intercept: This is the y-coordinate of the point where the graph crosses the y-axis.

Step 4: Find the values of y for several different values of x.

Step 5: Plot the points found in steps 2, 3, and 4.

Step 6: Draw the graph: Draw a smooth curve to connect the points.

Step 7: Identify any holes, or points where the graph is not defined, which occur at the x-coordinates where the denominator is zero and the numerator is not.

It's important to note that the degree of the numerator should be less than the degree of the denominator for the function to be defined for all x, except for the values that make the denominator zero. Also, the asymptotes of the graph are the vertical lines corresponding to the roots of the denominator.

Note: Rational functions can have some interesting properties, such as slant asymptotes, which occur when the degree of the denominator

is one less than the degree of the numerator, and cusp, which occur when the numerator and denominator share a common root.

Let's review an example:

Graph the rational function: $f(x) = \frac{x^2-x+2}{x-1}$.

First, notice that the graph is in two pieces. Most rational functions have graphs in multiple pieces. Since the polynomial of the numerator of the function does not have a real root, therefore the original function is also rootless.

Find the $y-$intercept by substituting zero for x and solving for y, $f(x)$:

$$x = 0 \to y = \frac{x^2-x+2}{x-1} = \frac{0^2-0+2}{0-1} = -2,$$

the $y-$intercept is the point of the coordinate $(0, -2)$.

Find the asymptotes of $\frac{x^2-x+2}{x-1}$:

Find the zeros of the function for the vertical asymptote as: $x - 1 = 0 \to x = 1$,

Divide the numerator by the denominator to find the slant asymptote: $\frac{x^2-x+2}{x-1} = x + \frac{2}{x-1}$.

So, the slant asymptote is the line $y = x$ (Exactly the divisor of this division).

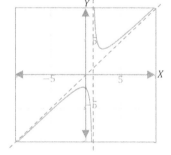

After finding the asymptotes, you can plug in some values for x and solve for y.

Here is the sketch for this function.

Note: When drawing the graph, make sure that the function graph does not intersect the asymptotes.

✎ Graph rational expressions.

11) $y = -\frac{1}{x}$

12) $y = \frac{x^2-4x+3}{x+1}$

13) $f(x) = \frac{x^2-2x}{x-3}$

14) $f(x) = \frac{6x+1}{x^2-4x}$

Adding and Subtracting Rational Expressions

When adding or subtracting rational expressions, it's important to first find a common denominator. A common denominator is a denominator that both rational expressions can be expressed with. Once a common denominator is found, you can add or subtract the numerators of the rational expressions, while keeping the common denominator.

Here's an example:

Solve: $\frac{3x+2}{x^2-1} + \frac{5x-4}{x^2-1} =$.

To add the rational expressions $\frac{3x+2}{x^2-1}$ and $\frac{5x-4}{x^2-1}$, we first find a common denominator which is $x^2 - 1$.

We can then add the numerators of the rational expressions while keeping the common denominator: $(3x + 2) + (5x - 4) = 8x\text{-}2$.

So, the sum of the rational expressions is $\frac{8x-2}{x^2-1}$.

$$\frac{3x+2}{x^2-1} + \frac{5x-4}{x^2-1} = \frac{(3x+2)+(5x-4)}{x^2-1} = \frac{8x-2}{x^2-1}.$$

Another example:

Solve: $\frac{3x+2}{x^2-1} - \frac{5x-4}{x^2-1} =$.

To subtract the rational expressions $\frac{3x+2}{x^2-1}$ and $\frac{5x-4}{x^2-1}$, we again use the common denominator $x^2 - 1$: $(3x + 2) - (5x - 4) = 3x + 2 - 5x + 4 = -2x + 6$.

So, the difference of the rational expressions is $\frac{-2x+6}{x^2-1}$.

$$\frac{3x+2}{x^2-1} - \frac{5x-4}{x^2-1} = \frac{(3x+2)-(5x-4)}{x^2-1} = \frac{-2x+6}{x^2-1}.$$

It's important to simplify the resulting expression after adding or subtracting, by canceling out common factors between numerator and denominator or using other methods of simplification.

It's also important to note that when the denominators are not the same, we need to find a common denominator to add or subtract the

Rational Expressions

expressions. Also, when the numerator and denominator share a common factor, we should cancel it out to get the simplified form of the expression.

Let's review an example:

Solve: $\frac{x+4}{x-5} + \frac{x-4}{x+6} =$.

Find the least common denominator of $(x-5)$ and $(x+6)$:

$$(x-5)(x+6) = x^2 + x - 30.$$

Then: $\frac{x+4}{x-5} + \frac{x-4}{x+6} = \frac{(x+4)(x+6)}{(x-5)(x+6)} + \frac{(x-4)(x-5)}{(x+6)(x-5)} = \frac{(x+4)(x+6)+(x-4)(x-5)}{(x+6)(x-5)}.$

Expand: $(x+4)(x+6) + (x-4)(x-5) = 2x^2 + x + 44.$

Then: $\frac{(x+4)(x+6)+(x-4)(x-5)}{(x+6)(x-5)} = \frac{2x^2+x+44}{(x+6)(x-5)} = \frac{2x^2+x+44}{x^2+x-30}.$

One more example:

Solve and simplify: $\frac{4}{2x+3} + \frac{x-2}{2x^2+x-3} =$.

To solve the equation, we need to first find a common denominator, which is in this case $2x^2 + x - 3$. Because $2x^2 + x - 3$ can be written as the product of two factors $2x + 3$ and $x - 1$, in which the denominator of the first part of the original expression is also present $2x + 3$. So,

$$\frac{4}{2x+3} + \frac{x-2}{2x^2+x-3} = \frac{4(x-1)}{2x^2+x-3} + \frac{x-2}{2x^2+x-3}$$

$$= \frac{4x-4+x-2}{2x^2+x-3}$$

$$= \frac{5x-6}{2x^2+x-3}.$$

✎ Simplify each expression.

15) $\frac{x+6}{x+1} - \frac{x+9}{x+1} =$

16) $\frac{2x+1}{x+3} + \frac{2}{x+4} =$

17) $\frac{14}{x+4} + \frac{6}{x^2-16} =$

18) $\frac{2}{x} + \frac{x-1}{x+1} - 1 =$

19) $\frac{x+2}{x+8} - \frac{2x}{x-8} =$

20) $\frac{x}{3x^2-x+1} + \frac{1}{3x^2-x+1} =$

Multiplying Rational Expressions

When multiplying rational expressions, we can multiply the numerators and denominators, separately. For example, to multiply the rational expressions $\frac{2x+3}{x-1}$ and $\frac{x+4}{x+2}$, we can use the following steps:

Step 1: Multiply the numerators:

$$(2x + 3) \times (x + 4) = 2x^2 + 8x + 3x + 12 = 2x^2 + 11x + 12.$$

Step 2: Multiply the denominators:

$$(x - 1) \times (x + 2) = x^2 + 2x - x - 2 = x^2 + x - 2.$$

Step 3: Write the product as the numerator over the denominator:

$$\frac{2x+3}{x-1} \times \frac{x+4}{x+2} = \frac{2x^2+11x+12}{x^2+x-2}.$$

Let's review an example:

Solve: $\frac{1}{x+3} \times \frac{x^2+x-2}{3x} =$.

Since the numerator and denominator polynomials of the fraction do not have a common factor, the product of this multiplication is obtained by multiplying the numerators and denominators, separately. So,

Multiply the numerators: $1 \times (x^2 + x - 2) = x^2 + x - 2$.

Multiply the denominators: $(x + 3) \times (3x) = 3x^2 + 9x$.

Write the product as the numerator over the denominator:

$$\frac{1}{x+3} \times \frac{x^2+x-2}{3x} = \frac{x^2+x-2}{3x^2+9x}.$$

It's also important to simplify the resulting expression, if possible, by canceling out common factors between the numerator and denominator or using other methods of simplification.

It's important to note that when we multiply fractions, we multiply the numerators and denominators separately, and then we simplify the result if possible.

Let's review another example:

Rational Expressions

Solve: $\frac{x^2+x}{x-1} \times \frac{2-2x}{x^3-4x^2+4x} =$.

Check the numerator and denominator polynomials to find the common factor. For this purpose, rewrite each of the polynomials as a product of factors. Then:

$\frac{x^2+x}{x-1} \times \frac{2-2x}{x^3-4x^2+4x} = \frac{x(x+1)}{x-1} \times \frac{-2(x-1)}{x(x-2)^2}$.

Now, cancel the common factors: $\frac{x(x+1)}{x-1} \times \frac{-2(x-1)}{x(x-2)^2} = (x+1) \times \frac{-2}{(x-2)^2}$.

Next, multiply the numerator of the fractions together and the denominator of the fractions together: $\frac{x^2+x}{x-1} \times \frac{2-2x}{x^3-4x^2+4x} = (x+1) \times \frac{-2}{(x-2)^2} = \frac{-2x-2}{(x-2)^2}$.

One more example:

Solve: $\frac{x-2}{x+3} \times \frac{2x+6}{x-2} =$.

First, we need to multiply the numerators and denominators, separately:

Multiply the numerators: $(x-2) \times (2x+6)$.

Multiply the denominators: $(x+3) \times (x-2)$.

Write the product as the numerator over the denominator:

$\frac{x-2}{x+3} \times \frac{2x+6}{x-2} = \frac{(x-2) \times (2x+6)}{(x+3) \times (x-2)}$.

Next, cancel the common factor: $\frac{(x-2) \times (2x+6)}{(x+3) \times (x-2)} = \frac{2x+6}{x+3}$.

Factor $(2x+6) = 2(x+3)$. Then: $\frac{2x+6}{x+3} = \frac{2(x+3)}{x+3} = 2$.

Therefore, the product of the multiplication is, $\frac{x-2}{x+3} \times \frac{2x+6}{x-2} = 2$.

✎ Simplify.

21) $\frac{20x^3}{3} \times \frac{15}{4x} =$

22) $\frac{x+6}{4} \times \frac{16}{x+6} =$

23) $\frac{x+10}{4x} \times \frac{3x}{7x+70} =$

24) $\frac{x+8}{x+6} \times \frac{x-6}{4x+32} =$

25) $\frac{x^2+1}{x^2-x} \times x =$

26) $\frac{x^2-2x}{x^3-1} \times \frac{x^2-1}{2x} =$

EffortlessMath.com

Dividing Rational Expressions

When dividing rational expressions, we can use the property that division is the same as multiplication by the reciprocal.

For example, to divide the rational expressions $\frac{2x+3}{x-1}$ by $\frac{x+4}{x+2}$, we can use the following steps:

Step 1: Flip the second fraction $\frac{x+4}{x+2}$:
$$\frac{x+2}{x+4}$$

Step 2: Multiply the first fraction $\frac{2x+3}{x-1}$ by the reciprocal of the second fraction:
$$\frac{2x+3}{x-1} \times \frac{x+2}{x+4}$$

Step 3: Simplify the resulting expression if possible by canceling out common factors between the numerator and denominator or using other methods of simplification.

So, the division is:
$$\frac{\frac{2x+3}{x-1}}{\frac{x+4}{x+2}} = \frac{2x+3}{x-1} \times \frac{x+2}{x+4} = \frac{(2x+3)(x+2)}{(x-1)(x+4)} = \frac{2x^2+4x+3x+6}{x^2+4x-x-4} = \frac{2x^2+7x+6}{x^2+3x-4}.$$

It's important to note that when we divide fractions, we flip the second fraction, then we multiply the numerators and denominators, separately, and then we simplify the result if possible.

Also, it's important to check the solutions in the original equation to make sure that they satisfy the equation.

Solve: $\frac{5x}{x+3} \div \frac{x}{2x+6} =$.

To solve this, we can use the following steps:

Step 1: Flip the second fraction $\frac{x}{2x+6}$ as: $\frac{2x+6}{x}$.

Step 2: Multiply the first fraction $\frac{5x}{x+3}$ by the reciprocal of the second

Rational Expressions

fraction $\frac{2x+6}{x}$: $\frac{5x}{x+3} \times \frac{2x+6}{x}$.

Step 3: Simplify the resulting expression if possible, by canceling out common factors between the numerator and denominator or using other methods of simplification.

So, the division is: $\frac{5x}{x+3} \div \frac{x}{2x+6} = \frac{5x}{x+3} \times \frac{2x+6}{x} = \frac{5x(2x+6)}{(x+3)x}$.

Now, cancel the common factor: $\frac{5x(2x+6)}{(x+3)x} = 10$.

So, the division is: $\frac{5x}{x+3} \div \frac{x}{2x+6} = 10$.

Another example:

Simplify the fraction: $\frac{\frac{x^2+x-6}{3x-3}}{\frac{x+3}{x^2-1}} =$.

To simplify, rewrite the fraction in the form:

$\frac{\frac{x^2+x-6}{3x-3}}{\frac{x+3}{x^2-1}} = \frac{x^2+x-6}{3x-3} \div \frac{x+3}{x^2-1}$.

Now, flip the second fraction and multiply by the first fraction as follows:

$\frac{x^2+x-6}{3x-3} \div \frac{x+3}{x^2-1} = \frac{x^2+x-6}{3x-3} \times \frac{x^2-1}{x+3}$.

Next, by finding the common factor from the denominator and numerator of the fraction, cancel the common factors and multiply the remaining expressions:

$\frac{x^2+x-6}{3x-3} \times \frac{x^2-1}{x+3} = \frac{(x-2)(x+3)(x-1)(x+1)}{3(x-1)(x+3)} = \frac{(x-2)(x+1)}{3} = \frac{x^2-x-2}{3}$.

✍ Simplify.

27) $\frac{10x}{x+2} \div \frac{x}{60x+120} =$

28) $\frac{5}{4} \div \frac{45}{8x} =$

29) $\frac{x-6}{x+3} \div \frac{4}{x+3} =$

30) $\frac{7x^3}{x^2-64} \div \frac{x^3}{x^2+x-56} =$

EffortlessMath.com

Evaluate Integers Raised to Rational Exponents

Integers raised to rational exponents can be evaluated using the properties of exponents. A rational exponent is a fraction with a numerator and denominator that are integers.

For example, to evaluate $2^{\frac{3}{4}}$, we can use the following steps:

- Step 1: Rewrite the exponent as a fraction with a numerator and denominator. In this case, $\frac{3}{4}$ is already in fraction form.
- Step 2: Write the base number (in this case, 2) as the numerator of a fraction with a denominator of 1.
- Step 3: Raise the base number to the power of the numerator of the exponent, and take the reciprocal of the base number to the power of the denominator of the exponent.

So, in this case: $2^{\frac{3}{4}} = (2^3)^{\frac{1}{4}} = 8^{\frac{1}{4}}$.

The result of this is the fourth root of 8.

It's important to note that these steps are valid as long as the denominator of the exponent is a positive integer, and the base is a positive real number.

Also, it's important to note that these steps are valid for integers, as well as real numbers.

If the base is a negative number, the exponent should be an even number, otherwise, the result will be a complex number.

Let's review a few examples:

Simplify: $(-32)^{-\frac{1}{5}}$.

To simplify $(-32)^{-\frac{1}{5}}$, we can use the properties of exponents. The negative exponent means that we need to find the reciprocal of the original expression. That is, use this formula $a^{-b} = \frac{1}{a^b}$, so: $(-32)^{-\frac{1}{5}} = \frac{1}{(-32)^{\frac{1}{5}}}$.

Now, we need to find the fifth root of 32. Write the base in exponential notation, $32 = 2^5$. It's important to note that the base number -32 is negative and the exponent is a rational number, this means that the result will be a complex number. But we know that the expression inside the radical with an odd index can also be negative and $\sqrt[m]{a} = a^{\frac{1}{m}}$. That is, the value $(-32)^{\frac{1}{5}}$ is defined and is a real value. Then:

$$(-32)^{\frac{1}{5}} = \left((-1) \times (32)\right)^{\frac{1}{5}} = ((-1)^5 \times 2^5)^{\frac{1}{5}} = (-1)^{\frac{5}{5}} \times 2^{\frac{5}{5}} = -1 \times 2 = -2.$$

Now, we can find the reciprocal of this expression: $(-32)^{-\frac{1}{5}} = \frac{1}{(-32)^{\frac{1}{5}}} = \frac{1}{-2} = -\frac{1}{2}$.

Evaluate: $81^{\frac{3}{2}}$.

First, rewrite the base as exponential form $81 = 3^4$. Now, we have: $81^{\frac{3}{2}} = (3^4)^{\frac{3}{2}}$. So, multiply two exponents, $(3^4)^{\frac{3}{2}} = 3^{4 \times \frac{3}{2}} = 3^6 = 729$.

Calculate the value of $\sqrt[3]{121^{\frac{3}{2}}}$.

Use this formula $\sqrt[n]{a} = a^{\frac{1}{n}}$. Then: $\sqrt[3]{121^{\frac{3}{2}}} = \left(121^{\frac{3}{2}}\right)^{\frac{1}{3}}$.

Now, by using $(a^n)^m = a^{nm}$. So, we have: $\left(121^{\frac{3}{2}}\right)^{\frac{1}{3}} = 121^{\frac{3}{2} \times \frac{1}{3}} = 121^{\frac{1}{2}}$.

Write the base in exponential notation, $121^{\frac{1}{2}} = (11^2)^{\frac{1}{2}}$.

Multiply the exponents and simplify: $(11^2)^{\frac{1}{2}} = 11^{2 \times \frac{1}{2}} = 11$.

✎ Evaluate.

31) $(121x^6)^{\frac{1}{2}}$

32) $(64x^{12})^{\frac{1}{6}}$

33) $(32)^{-\frac{1}{5}}$

34) $(-27)^{\frac{1}{3}}$

35) $\left(\sqrt{256}a^2\right)^{-\frac{3}{2}}$

36) $\sqrt{64^{\frac{5}{3}}}$

Chapter 10: Answers

1) $\frac{3}{19}$

2) $-\frac{60}{7}$

3) $\frac{38}{67}$

4) $-\frac{15}{58}$

5) $\frac{3}{32}$

6) $\frac{35}{36}$

7) $\frac{20}{3}$

8) $\frac{11x}{6}$

9) $\frac{2x+6}{3x-6}$

10) $\frac{4+x}{4x}$

11) $y = -\frac{1}{x}$

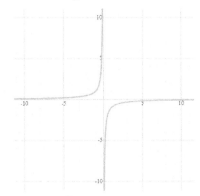

12) $y = \frac{x^2-4x+3}{x+1}$

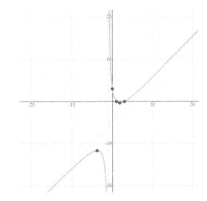

13) $f(x) = \frac{x^2-2x}{x-3}$

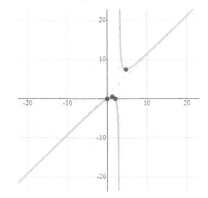

14) $f(x) = \frac{6x+1}{x^2-4x}$

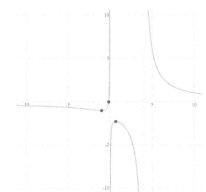

15) $-\frac{3}{x+1}$

16) $\frac{2x^2+11x+10}{(x+3)(x+4)}$

17) $\frac{14x-50}{(x+4)(x-4)}$

18) $\frac{2}{x^2+x}$

19) $\frac{-x^2-22x-16}{(x+8)(x-8)}$

20) $\frac{x+1}{3x^2-x+1}$

21) $25x^2$

22) 4

Rational Expressions

23) $\frac{3}{28}$

24) $\frac{x-6}{4(x+6)}$

25) $\frac{x^2+1}{x-1}$

26) $\frac{(x-2)(x+1)}{2(x^2+x+1)}$

27) 600

28) $\frac{2x}{9}$

29) $\frac{x-6}{4}$

30) $\frac{7(x-7)}{x-8}$

31) $11x^3$

32) $2x^2$

33) $\frac{1}{2}$

34) -3

35) $\frac{1}{2^6 a^3}$

36) 32

CHAPTER 11: Statistics and Probabilities

Math topics that you'll learn in this chapter:

☑ Mean, Median, Mode, and Range of the Given Data
☑ Pie Graph
☑ Scatter Plots
☑ Probability Problems
☑ Permutations and Combinations
☑ Calculate and Interpret Correlation Coefficients
☑ Find the Equation of a Regression Line and Interpret Regression Lines
☑ Correlation and Causation

Mean, Median, Mode, and Range of the Given Data

Mean, median, mode, and range are four common measures of central tendency and variability in statistics.

Mean: The mean is the average of a set of data. It is calculated by adding all the values in the set and dividing the sum by the number of values. The formula for the mean is:

$$mean = \frac{sum\ of\ values}{number\ of\ values}$$

Median: The median is the middle value when a set of data is arranged in numerical order. If there is an even number of values, the median is the average of the two middle values.

Mode: The mode is the most frequently occurring value in a set of data. A set of data can have one mode, more than one mode, or no mode at all.

Range: The range is the difference between the highest and lowest values in a set of data. It is a measure of variability in the set. The formula for the range is:

$$range = highest\ value - lowest\ value$$

It's important to note that these measures are not always sufficient to describe the data, and in some cases, we may need to use other measures such as variance, standard deviation, etc. Also, it's important to note that the data should be numerical to be able to calculate these measures.

Let's review an example:

Find the mean, median, mode, and range of the following numbers:

$$5, 6, 8, 6, 8, 5, 3, 6$$

Mean: To find the mean, add up all the values and divide by the number of values. Therefore, the mean is equal to $\frac{5+6+8+6+8+5+3+6}{8} = 5.87$.

Median: To find the median, arrange the values in numerical order and find the middle value. 3, 5, 5, 6, 6, 6, 8, 8

In this case, since there is an even number of values, the median is the average of the two middle values: $\frac{6+6}{2} = 6$.

Mode: To find the mode, find the most frequently occurring value. In this case, the mode is 6, which occurs 3 times.

Range: To find the range, subtract the lowest value from the highest value: $8 - 3 = 5$.

Some other examples:

What is the mode of these numbers? 5, 6, 8, 6, 8, 5, 3, 5

Mode: the value in the list that appears most often.

Therefore, the mode is number 5. There are three number 5 in the data.

What is the median of these numbers? 6, 11, 15, 10, 17, 20, 7

Write the numbers in order: 6, 7, 10, 11, 15, 17, 20

The median is the number in the middle. Therefore, the median is 11.

What is the mean of these numbers? 7, 2, 3, 2, 4, 8, 7, 5

Use this formula: Mean: $\frac{sum\ of\ values}{number\ of\ values}$.

Therefore: Mean: $\frac{7+2+3+2+4+8+7+5}{8} = \frac{38}{8} = 4.75$.

What is the range in this list? 3, 7, 12, 6, 15, 20, 8

The range is the difference between the largest value and the smallest value in the list. The largest value is 20 and the smallest value is 3. Then: $20 - 3 = 17$.

✒ Find the mean, median, mode, and range of the following.

1) 6, 11, 5, 3, 6

2) 4, 9, 1, 9, 6, 7

3) 10, 3, 6, 10, 4, 15

4) 12, 4, 8, 9, 3, 12, 15

Pie Graph

A pie chart, also known as a pie graph or a circle graph, is a type of chart used to represent data as a proportion of the whole. It is useful for showing the relative sizes of different categories or parts of a whole.

A pie chart is made up of a circular area divided into wedges, with each wedge representing a proportion of the whole. The size of each wedge is determined by the proportion of the category it represents to the whole. The wedges are usually labeled with the category name and the percentage of the whole that the category represents.

To create a pie chart, you will need data that can be divided into categories or parts. Each category or part will be represented by a wedge in the chart. The total of the categories should equal 100%.

You can create a pie chart using charting software, spreadsheet software, or an online chart-making tool.

It's important to note that a pie chart is best used when there are few categories (less than 6) and the categories are not too close in size, otherwise, it's hard to compare the sizes of the wedges.

Let's review an example:

A library has 750 books that include Mathematics, Physics, Chemistry, English and History. Use the following graph to answer the questions.

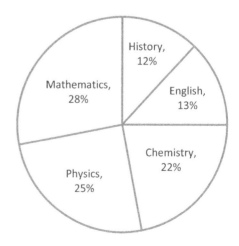

What is the number of Mathematics books?

Number of total books = 750.

Statistics and Probabilities

Percent of Mathematics books = 28%.

Then, the number of Mathematics books: 28% × 750 = 0.28 × 750 = 210.

What is the number of History books?

Number of total books = 750.

Percent of History books = 12%.

Then: 0.12 × 750 = 90.

What is the number of Chemistry books in the library?

Number of total books = 750.

Percent of Chemistry books = 22%.

Then: 0.22 × 750 = 165.

✎ **The pie chart shown below shows the percentages of types of transportation used by 800 students to come to school. With this given information, answer the following questions:**

5) How much did Bob's total expenses last month? _____

6) How much did Bob spend on food last month? _____

7) How much did Bob spend on his bills last month? _____

8) How much did Bob spend on his car last month? _____

Bpb's last month expenses

Scatter Plots

A scatter plot, also known as a scatter graph or scatter chart, is a type of plot used to show the relationship between two quantitative variables. It is a way to represent the data in the form of points on a two-dimensional graph, with one variable on the x-axis and the other variable on the y-axis. Each point on the scatter plot represents a single observation in the data set, with the x-coordinate representing the value of one variable and the y-coordinate representing the value of the other variable.

Scatter plots are useful for visualizing patterns in data, such as correlation, clustering, and outliers. A scatter plot can show a positive correlation, a negative correlation, or no correlation at all between the two variables. If the points on the plot tend to fall along a straight line, then there is likely a strong correlation between the two variables.

Scatter plots can be created using spreadsheet software, charting software, or online chart-making tools. When creating a scatter plot, it is important to label the axes and provide a title that describes the variables being plotted.

In some cases, Scatter plots can be useful to detect outliers, which are data points that are significantly different from the other points in the dataset.

It's important to note that scatter plots are best used when there is more than one observation, and the data is numerical.

Let's review an example:

The following table shows the number of people in a family and the amount of money they spend on movie tickets.

Number of People	1	2	3	4	5	6	7
Money ($)	13	14	17	15	28	18	16

Statistics and Probabilities

a) Make a scatter plot to represent the data.

b) Does this scatter plot show a positive trend, a negative trend, or no trend?

c) Find the outlier on the scatter plot.

Write the ordered pairs. Number of people goes on the x-axis, so put the number of people first. The amount of money goes on the y-axis, so put the amount of money second. (1,13), (2,14), (3,17), (4,15), (5,28), (6,18), (7,16).

Now, graph the ordered pairs.

y tends to increase as x increases. So, this scatter plot shows a positive trend.

(5,28) is the outlier because this point is separated from all other points in the data set.

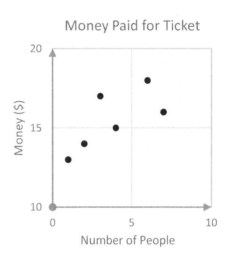

✎ **In the scatterplots below, determine which is a positive trend, negative or no trend, and find the outlier if there is one.**

9)

Day	Money ($)
1	$12
2	$8
3	$14
4	$5
5	$7
6	$10
7	$4
8	$12

10)

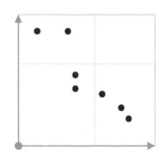

Probability Problems

Probability problems involve using mathematical concepts and techniques to determine the likelihood of certain events occurring. These types of problems can involve a wide range of scenarios, from simple coin flips and dice rolls to more complex real-world situations.

For example, if you flip a coin 10 times and it lands on heads 6 times, the experimental probability of flipping heads on the next flip would be $\frac{6}{10}$ or $\frac{3}{5}$.

Probability theory provides the foundation for many concepts and techniques used in data analysis and decision-making, and it is an important area of study for anyone interested in the mathematical and scientific foundations of these fields.

There are several ways to solve probability problems, but the general process involves the following steps:

Step 1: Define the sample space: The sample space is the set of all possible outcomes of the experiment or event being considered.

Step 2: Define the event: The event is the specific outcome or set of outcomes that we are interested in finding the probability of.

Step 3: Use the formula for probability: The probability of an event is the number of favorable outcomes (the outcomes that make up the event) divided by the total number of possible outcomes in the sample space. This is represented by the formula: $P(E) = \frac{n(E)}{n(S)}$, where $P(E)$ is the probability of the event E, $n(E)$ is the number of favorable outcomes and, $n(S)$ is the total number of possible outcomes in the sample space S.

Step 4: Check for the probability of the event to be: $0 \leq P(E) \leq 1$.

Step 5: Find the complementary probability if necessary. The complementary probability is the probability that the event will not happen and it's represented by the formula $P(E') = 1 - P(E)$.

Statistics and Probabilities

It's important to note that the sample space should be exhaustive, meaning that should include all possible outcomes, and mutually exclusive, meaning that the sample space should not contain overlapping outcomes.

Probability problems can be tricky, but with practice and understanding of the concepts, you'll be able to solve them with ease.

Probability can be expressed as a fraction, a decimal, or a percent.

Let's review an example:

Anita's trick–or–treat bag contains 10 pieces of chocolate, 16 suckers, 16 pieces of gum, 22 pieces of licorice. If she randomly pulls a piece of candy from her bag, what is the probability of her pulling out a piece of sucker?

Use this formula: $P(E) = \frac{n(E)}{n(S)} = \frac{number\ of\ favorable\ outcomes}{the\ total\ number\ of\ possible\ outcomes}$.

Probability of pulling out a piece of sucker $= \frac{16}{10+16+16+22} = \frac{16}{64} = \frac{1}{4}$.

One more example:

A bag contains 20 balls: four green, five black, eight blue, a brown, a red and one white. If 19 balls are removed from the bag at random, what is the probability that a brown ball has been removed?

If 19 balls are removed from the bag at random, there will be one ball in the bag. The probability of choosing a brown ball is 1 out of 20. Therefore, the probability of not choosing a brown ball is 19 out of 20 and the probability of having not a brown ball after removing 19 balls is the same. The answer is: $\frac{19}{20}$.

✎ Solve.

11) Bag A contains 8 red marbles and 6 green marbles. Bag B contains 5 black marbles and 7 orange marbles. What is the probability of selecting a green marble at random from bag A? What is the probability of selecting a black marble at random from Bag B?

Permutations and Combinations

Permutations and combinations are two related concepts in combinatorics, the branch of mathematics that deals with counting and arranging objects in different ways.

Permutations refer to the number of ways in which a set of objects can be arranged in a specific order. For example, if you have 3 objects: *A*, *B*, and *C*, there are 3! (3 factorial) or 6 possible permutations: *ABC*, *ACB*, *BAC*, *BCA*, *CAB*, *CBA*.

Combinations, on the other hand, refer to the number of ways in which a set of objects can be chosen without regard to order. For example, if you have 3 objects: *A*, *B*, and *C*, there are 3 choose 2 or 3 possible combinations: *AB*, *AC*, *BC*.

The formula for permutations of n objects taken r at a time is $\frac{n!}{(n-r)!}$, and the formula for combinations of n objects taken r at a time is $nCr = \frac{n!}{r!(n-r)!}$.

Permutations and combinations are used in a variety of fields such as combinatorial game theory, coding theory, cryptography, statistics, and probability.

It's important to note that the order of the elements matter when it comes to permutations while it doesn't matter in combinations.

In addition, Permutations are used when the order of elements matters, like arranging people in a line, while combinations are used when the order doesn't matter, like choosing a committee. In summary:

- **Factorials** are products, indicated by an exclamation mark. For example, $4! = 4 \times 3 \times 2 \times 1$. (Remember that 0! Is defined to be equal to 1.)
- **Permutations:** The number of ways to choose a sample of k elements from a set of n distinct objects where order does matter, and replacements are not allowed. For a permutation problem, use this formula: $nPk = \frac{n!}{(n-k)!}$.

- **Combination:** The number of ways to choose a sample of r elements from a set of n distinct objects where order does not matter, and replacements are not allowed. For a combination problem, use this formula: $nCr = \frac{n!}{r!(n-r)!}$.

Let's review an example:

How many ways can the first and second place be awarded to 7 people?

Since the order matters, (The first and second places are different!) we need to use permutation formula where n is 7 and k is 2. Then: $\frac{n!}{(n-k)!} = \frac{7!}{(7-2)!} = \frac{7!}{5!} = \frac{7 \times 6 \times 5!}{5!}$, remove 5! From both sides of the fraction. Then: $\frac{7 \times 6 \times 5!}{5!} = 7 \times 6 = 42$.

How many ways can we pick a team of 3 people from a group of 8?

Since the order doesn't matter, we need to use a combination formula where n is 8 and r is 3. Then: $\frac{n!}{r!(n-r)!} = \frac{8!}{3!(8-3)!} = \frac{8!}{3!(5)!} = \frac{8 \times 7 \times 6 \times 5!}{3!(5)!} = \frac{8 \times 7 \times 6}{3 \times 2 \times 1} = \frac{336}{6} = 56$.

✎ Determine whether each situation involves a permutation or a combination.

12) Susan is baking cookies. She uses sugar, flour, butter, and eggs. How many different orders of ingredients can she try?

13) Jason is planning for his vacation. He wants to go to a museum, go to the beach, and play volleyball. How many different ways of ordering are there for him?

14) In how many ways can a team of 6 basketball players choose a captain and co-captain?

15) How many ways can you give 5 balls to your 8 friends?

16) A professor is going to arrange her 5 students in a straight line. In how many ways can she do this?

17) In how many ways can a teacher choose 12 out of 15 students?

Calculate and Interpret Correlation Coefficients

A correlation coefficient is a statistical measure that describes the strength and direction of a linear relationship between two variables. The correlation coefficient is used for a set of n data points, (x_i, y_i), where $1 \leq i \leq n$. The most commonly used correlation coefficient is the Pearson correlation coefficient, denoted by r.

The Pearson correlation coefficient can range from -1 to 1, where -1 indicates a perfect negative linear relationship, 0 indicates no linear relationship, and 1 indicates a perfect positive linear relationship. A positive value of r indicates that as one variable increases, the other variable also increases, while a negative value of r indicates that as one variable increases, the other variable decreases. The formula for the Pearson correlation coefficient is:

$$r = \frac{\sum(x_i - \bar{x})(y_i - \bar{y})}{\sqrt{\sum(x_i - \bar{x})^2 \sum(y_i - \bar{y})^2}}$$

where x_i and y_i are the values of the two variables for the i-th observation, n is the total number of observations, $\sum$ (Sigma) denotes the sum of the values, and $\bar{x}$ and $\bar{y}$ are the mean of the variables.

A correlation coefficient of 0.8 or higher is considered strong, a coefficient between 0.5 and 0.8 is moderate, and a coefficient below 0.5 is weak. However, it's important to note that correlation doesn't imply causation, it only implies a relationship.

The correlation coefficient is a useful tool for identifying patterns in data, but it is important to remember that it does not indicate causality and should not be used to infer cause-and-effect relationships between variables. Also, it's important to note that the correlation coefficient only measures linear relationships, so if there's a non-linear relationship between variables, the correlation coefficient would not be an appropriate measure.

bit.ly/3QWsz0c
Find more at

Let's review an example:

Find the correlation coefficient of the following data and then interpret the answer. Round your answer to the nearest thousandth.

Arthur plans to improve his studies. In order to achieve this goal, he has written down the number of pages read daily and the hours of daily reading in the table below. He records the hours of daily reading, x, and the number of pages read daily, y.

Hours of Daily Reading	Number of Pages Read Daily
3	60
4	75
6	82
7	90
8	105

Each row in the above table shows a data point (x_i, y_i). x_i is the number of hours Arthur reads per day and y_i is the number of pages Arthur reads per day. You can simply use your calculator to find the above data set's correlation coefficient: $r = 0.968971 \approx 0.969$. The correlation coefficient is positive. So, the data points have a positive trend or increasing trend. The correlation coefficient is so close to 1, So the data set has a strong linear correlation.

✎ Find the correlation coefficient of the following data.

18)

x	50	51	52	53	54
y	4.1	4.2	4.3	4.4	4.5

19)

x	12	14	18	21	28
y	2	4	6	8	12

Find the Equation of a Regression Line and Interpret Regression Lines

A regression line is a line that is used to model the relationship between two variables. The equation of a regression line is typically represented in the form of the equation $y = bx + a$, where y is the dependent, x is the independent variable, a is the y-intercept (the point at which the line crosses the y-axis), and b is the slope of the line (the rate of change of y with respect to x).

The slope of line, b, can be calculated using the following formula:

$$b = \frac{n \sum (x_i y_i) - \sum x_i \sum y_i}{n \sum x_i^2 - (\sum x_i)^2}$$

The y-intercept, a, can be calculated using the following formula:

$$a = \frac{\sum y_i - b \sum x_i}{n}$$

where x_i and y_i are the values of the two variables for the i-th observation, n is the total number of observations, and $\sum$ (Sigma) denotes the sum of the values.

Once the slope and y-intercept are calculated, the equation of the regression line can be written as:

$$y = a + bx$$

The slope of the line, b, can be interpreted as the rate of change of y with respect to x, and the y-intercept, a, can be interpreted as the expected value of y when $x = 0$.

The equation of the regression line can be used to make predictions about the value of the dependent variable (y) based on the value of the independent variable (x). For example, if the equation of the regression line is $y = 2 + 3x$, then a value of $x = 2$ would correspond to a predicted value of $y = 8$.

It's important to note that a regression line is a model, and it's always going to have some error. The error can be measured by the coefficient

of determination (R^2), which is a number between 0 and 1 that indicates how well the line fits the data. An $R^2 = 0$, indicates that the line fits the data perfectly, while an $R^2 = 1$, indicates that the line does not fit the data at all.

$$y = a + bx + R^2$$

Let's review an example:

Find the equation for the least squares regression line of the following data set. Round your answers to the nearest hundredth. Then interpret the regression line.

Use a calculator to find the least squares regression line's equation for the above data set: $y = -0.15875x + 22.53709$. Round your answers to the nearest hundredth:

$$y = -0.16x + 22.54.$$

x_i	y_i
9	24
15	20
23	17
32	13
45	9

The slope is negative so, there is a negative linear relationship. It means when one variable increases the other variable decreases.

✍ Determine the linear regression equation from the given set of data.

20)

x	2	3	5	8
y	3	6	4	13

21)

x	2	4	6	8
y	4	7	10	12

Correlation and Causation

Correlation and causation are two related but distinct concepts in statistics. Correlation refers to a statistical relationship between two variables, where the values of one variable tend to change in a predictable way in relation to the values of the other variable. A correlation coefficient (such as Pearson's r) is a numerical measure of the strength and direction of this relationship.

Causation, on the other hand, refers to the relationship between an event (the cause) and a second event (the effect), where the second event is a result of the first event. In other words, causation implies that one variable causes the other variable to change.

It's important to note that correlation does not imply causation. Just because two variables are correlated, it does not mean that one variable causes the other. For example, ice cream sales and crime rates are positively correlated, but it doesn't mean that ice cream causes crime. The reason for the correlation is that both ice cream sales and crime rates increase during the summer months.

To establish causation, it's important to use experimental methods, such as randomized controlled trials, where the effect of the independent variable on the dependent variable can be established. Additionally, it's important to consider other factors that might be influencing the relationship between the two variables and controlling for any potential confounding variables.

One example of correlation not implying causation is the relationship between the number of fire trucks in a city and the number of fires in that city. There is likely a positive correlation between the number of fire trucks and the number of fires - as the number of fire trucks increases, the number of fires

also increases. However, it doesn't mean that having more fire trucks causes more fires.

The reason for the correlation is that both the number of fire trucks and the number of fires are related to the population size of the city. As the population of the city grows, the number of fire trucks and the number of fires also increases. Therefore, we cannot conclude that the number of fire trucks causes the number of fires, instead, the relationship is due to another variable (population size) that influences both.

It's important to note that correlation is just a statistical relationship between two variables and it should be interpreted with caution. It's necessary to use other methods to establish the causal relationship between variables.

Keep in mind a correlation does not signify causation, but causation always signifies correlation. There are 2 situations when a correlation isn't causation:

The third variable problem happens when a confounded variable affects two other variables and makes them seem causally linked when they are not.

The directionality problem happens when 2 variables have a correlation connection and might really have a causal link, but it isn't possible to determine which variable is the reason for changing in the other.

✍ Determine whether the following relationships reflect both correlation and causation or not.

22) The number of cold and snowy days and the amount of coffee at the ski resort.

23) The number of miles traveled and the gas used.

Chapter 11: Answers

1) Mode: 6, Range: 8, Mean: 6.2, Median: 6

2) Mode: 9, Range: 8, Mean: 6, Median: 6.5

3) Mode: 10, Range: 12, Mean: 8, Median: 8

4) Mode: 12, Range: 12, Mean: 9, Median: 9

5) $1,975

6) $158

7) $730.75

8) $197.5

9) No trend

10) Negative trend

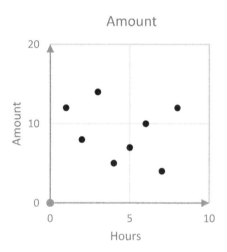

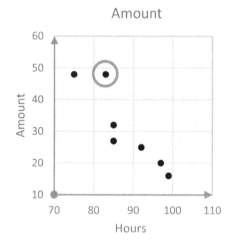

11) $\frac{3}{7}, \frac{5}{12}$

12) 24 (It's a permutation problem.)

13) 6 (It's a permutation problem.)

14) 30 (It's a permutation problem.)

15) 792 (It's a combination problem.)

16) 120 (It's a permutation problem.)

17) 455 (It's a combination problem.)

18) 1

19) 0.9971

20) $y = 1.476x - 0.142$

21) $y = 1.35x + 1.5$

22) Correlation no causation

23) Correlation and causation

Chapter 12: Direct and Inverse Variation

Math topics that you'll learn in this chapter:

☑ Find the Constant of Variation
☑ Model Inverse Variation
☑ Write and Solve Direct Variation Equations

Find the Constant of Variation

A constant of variation is a relationship between two variables that can be expressed as a constant multiplied by one of the variables. The constant of variation is also known as the proportionality constant or the coefficient of proportionality.

Constant variation is usually seen in one of these two types of formats: direct variation or inverse variation. In both of these formats, k is considered the constant of variation. If you have an equation in one of these two formats, you can easily find the constant of variation.

$$\text{Direct Variation} \to y = kx \qquad \text{Inverse Variation} \to y = \frac{k}{x}$$

Where k is a non-zero constant, x and y are variables.

The most common form of constant variation is direct variation, where the two variables y and x are directly proportional to each other. The value of k represents how much y changes for a unit change in x.

In contrast, inverse variation is also another form of constant variation, where the two variables y and x are inversely proportional. This equation shows that as the value of x increases, the value of y decreases and vice versa.

It's important to note that in both cases, the constant of variation is a fixed value that doesn't change, regardless of the specific values of x and y.

Note that all equations don't have a constant of variation and they can't be written this way.

An example of a direct variation is the relationship between distance and the time it takes to travel that distance at a certain speed. Let's say you're driving on a highway at a constant speed of 60 miles per hour. The distance you travel is directly proportional to the time it takes to travel that distance. If you drive for 1 hour, you will travel 60 miles ($1 \: hour \times 60 \frac{miles}{hour} = 60 \: miles$). If you drive for 2 hours, you will travel 120 miles (2 $hours \times 60 \frac{miles}{hour} = 120\ miles$). The equation that represents this relationship is: $Distance = Speed \times Time$.

In this case, the constant of variation (k) is equal to the speed of 60 miles per hour.

An example of an inverse variation is the relationship between the weight of an object and the gravitational force acting on it. The force of gravity acting on an object is inversely proportional to the distance between the object and the center of the Earth. The equation that represents this relationship is:

$$F = \frac{G(m_1 \times m_2)}{d^2}$$

Where F is the force of gravity, G is the gravitational constant, m_1 and m_2 are the masses of the objects and d is the distance between the objects.

In this case, the constant of variation k is equal to $G(m_1 \times m_2)$ and as the distance between the objects increases, the force of gravity decreases and vice versa.

One more example:

If y varies directly as x, and $x = 24$ when $y = 21$, what is the direct variation equation?

First, put x and y values in the *Direct Variation* $\rightarrow y = kx$ to find the constant of variation: $y = kx \rightarrow k = \frac{y}{x} = \frac{21}{24} \rightarrow k = \frac{7}{8}$.

Now using the constant of variation, write the direct variation equation: $y = \frac{7}{8}x$.

✑ Find the constant of variation in each case.

1) Let x and y be in direct variation, $x = 6$ and $y = 22$. Then find the direct variation equation.

2) If $x = 15$ and $y = 30$ follow a direct variation then find the constant of proportionality.

3) If y varies inversely as x, and $x = 5$ when $y = 7$, what is the inverse variation equation?

Model Inverse Variation

An inverse variation is a relationship between two variables which is shown by the following equation:

$$y = \frac{k}{x}$$

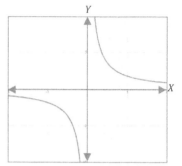

Sample Graph of Inverse Variation

Where k is a non-zero constant, x and y are variables. This equation shows that as the value of x increases, the value of y decreases, and vice versa.

To model inverse variation, you can use the equation to find the value of k and then use it to find the value of y for a given value of x or vice versa.

For example, let's say we have an inverse variation between the distance of a planet from the sun and its orbital period. The data shows that when a planet is 4 million miles from the sun, its orbital period is 365 days and when a planet is 40 million miles from the sun, its orbital period is 10,000 days. We can use this data to find k and then use it to find the orbital period of a planet that is 20 million miles from the sun.

$$y = \frac{k}{x} \rightarrow k = xy \rightarrow k = 365 \times 4 = 1460$$

So, the equation becomes: $y = \frac{1460}{x}$

To find the orbital period of a planet that is 20 million miles from the sun, we can substitute $x = 20$ million miles into the equation: $y = \frac{1460}{20} = 73$

So, the orbital period of a planet that is 20 million miles from the sun is 73 days.

In general, when modeling inverse variation, you can use any given pair of x and y values to find k, and then use the resulting equation to find the value of the other variable for a given value of x or y.

Direct and Inverse Variation

Another example:

According to the following table, determine if *y* values change inversely with *x*.

x	3	4	6	8
y	16	12	8	6

If yes, write an equation for the inverse variation and show it in a graph.

In a table with inverse changes, the product of all *x* and *y* pairs of its data are the same value. You can see that the product of any pair of *x* and *y* is equal to 48, so: $k = 48$. write an equation for the inverse variation: $y = \frac{k}{x} \rightarrow y = \frac{48}{x}$.

You can make a graph of the equation $y = \frac{48}{x}$ with points from the table as follows.

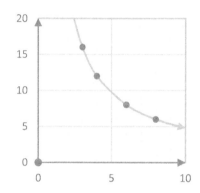

✍ In each case, state the constant of variation and graph the model.

4)

x	*y*
2	12
4	6
6	4
8	3

5)

x	*y*
2	15
3	10
5	6
6	5

Write and Solve Direct Variation Equations

Direct variation is a mathematical relationship that exists between two variables, where one variable is directly proportional to the other. In other words, as the value of one variable increases, the value of the other variable also increases in a consistent manner, and vice versa. The ratio between the two variables always remains the same in a direct variation.

The equation of a direct variation can be written as $y = kx$, where y and x are the two variables in question, and k is the constant of variation. The constant of variation is a value that remains the same throughout the equation and represents the ratio between the two variables.

To solve a direct variation equation, we can use the given values of the variables and the constant of variation to find the unknown variable.

For example, if we are given $y = 3x$ and $x = 4$, we can substitute the value of x into the equation to find the value of y: $y = 3x \rightarrow y = 3(4) \rightarrow y = 12$.

Therefore, the value of y in this example is 12.

Another example, if we are given the equation $y = 2x$ and $y = 7$. We can substitute the value of y into the equation and solve for x:

$$y = 2x \rightarrow 7 = 2x \rightarrow x = 3.5$$

Therefore, the value of x in this example is 2.

It's worth noting that the graph of a direct variation is always a straight line, with the constant of variation represented by the slope of the line. Considering the last issue, the following method can be used to write and solve the equation of a direct variation.

A two-variables linear equation's slope-intercept form is as follows: $y = mx + b$. Now, if you consider the y-intercept occurs at the point of (0,0) and substitute the slope (m) with the constant of variation (k), you get a direct variation

Direct and Inverse Variation

equation: $y = kx$, where k is the constant of variation and it's a value that always remains with no changes.

In the direct variation, the proportion between the variables stays the same, therefore the direct variation graph always makes a straight line.

Let's review some examples:

If y varies directly with x and $y = 21$ when $x = 7$, find y when $x = 8$.

First, you should find the constant of variation. Put $x = 3$ and $y = 21$ into the direct variation equation to find the value of k:

$$y = kx \to 21 = k \times 7 \to k = \frac{21}{7} = 3 \to k = 3.$$

Now, use $k = 3$ to find y when $x = 8$: $y = kx \to y = 3 \times 8 = 24 \to y = 24$.

If y varies directly with x and $y = 16$ when $x = 4$, find x when $y = 32$.

First, you should find the constant of variation. Put $x = 4$ and $y = 16$ into the direct variation equation to find the value of k:

$$y = kx \to 16 = k \times 4 \to k = \frac{16}{4} = 4 \to k = 4.$$

Now, use $k = 4$ to find x when $y = 32$: $y = kx \to 32 = 4 \times x = x = \frac{32}{4} = 8 \to x = 8$.

If y varies directly with x and $y = 42$ when $x = 6$, find y when $x = 11$.

First, you should find the constant of variation. Put $x = 6$ and $y = 42$ into the direct variation equation to find the value of k:

$$y = kx \to 42 = k \times 6 \to k = \frac{42}{6} = 7 \to k = 7.$$

Now, use $k = 7$ to find y when $x = 11$: $y = kx \to y = 7 \times 11 = 77 \to y = 77$.

✎ Solve.

6) If y varies directly with x and $y = 8$ when $x = 12$, find y when $x = -6$.

7) If y varies directly with x, find the missing value of x in $(-3, 27)$ and $(x, -27)$.

8) If y varies directly with x and $y = 12$ when $x = 2$, find y when $x = 7$.

Chapter 12: Answers

1) $y = \frac{11}{3}x$

2) $k = 2$

3) $y = \frac{35}{x}$

4) $y = \frac{24}{x}$

5) $y = \frac{30}{x}$

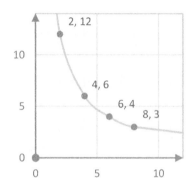
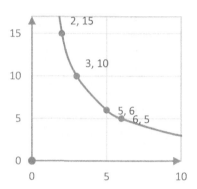

6) $y = -4$

7) $x = 3$

8) $y = 42$

CHAPTER 13
Number Sequences

Math topics that you'll learn in this chapter:

- ☑ Evaluate Recursive Formulas for Sequences
- ☑ Evaluate Variable Expressions for Number Sequences
- ☑ Write Variable Expressions for Arithmetic Sequences
- ☑ Write Variable Expressions for Geometric Sequences
- ☑ Write a Formula for a Recursive Sequence

Evaluate Recursive Formulas for Sequences

A sequence is an ordered list of numbers, where each number is called a term. The terms of a sequence are usually represented by the notation $T_1, T_2, \cdots, T_n, \cdots$, where T_n denotes the nth term of the sequence. The position of the term in the sequence is called an index and is represented by the subscript n. Sequences can be defined in different ways, such as an explicit formula, a recursion formula, or a set of given terms.

A recursive sequence is a sequence whose general term defines each term of the sequence using a previous term or terms. The general formula for a recursive sequence is $T_{n+1} = f(T_n)$, where T_n represents the nth term of the sequence, and f is a function that relates the nth term to one or more previous terms. To find the terms of a recursive sequence, it is necessary to know several previous terms of the sequence and use the recursive formula to find the next term.

Let's review an example:

Find a_2 and a_3, where the first term is $a_1 = -1$, and the general term is:

$$a_n = 2a_{n-1} - 1$$

Use the recursive sequence $a_n = 2a_{n-1} - 1$.

To find a_2, plug $n = 2$ in $a_n = 2a_{n-1} - 1$. So, $a_2 = 2a_{2-1} - 1 \rightarrow a_2 = 2a_1 - 1$. Now, substitute $a_1 = -1$ into the obtained formula $a_2 = 2a_1 - 1$.

Therefore, $a_2 = 2(-1) - 1 = -3$.

In the same way, plug in $n = 3$. Then, $a_3 = 2a_{3-1} - 1 \rightarrow a_3 = 2a_2 - 1$.

To find the value of a_3, put the obtained value for a_2 in this relation $a_3 = 2a_2 - 1$.

Therefore, $a_2 = -3 \rightarrow a_3 = 2(-3) - 1 = -7$.

The terms are $a_2 = -3$ and $a_3 = -7$.

Another example:

Number Sequences

Write the first five terms of the sequence $T_n = T_{n-1} + T_{n-2}$, where $n \geq 3$, $T_1 = 1$ and $T_2 = 3$.

Recall that the recursive formula defines each term of a sequence using a previous term or terms.

Then by looking at the recursive formula $T_{n+1} = T_n + T_{n-1}$, you notice that to generate the terms of the sequence, you need the previous two terms of the sequence.

Here you are given the first two terms $T_1 = 1$ and $T_2 = 3$ together with the recursive formula $T_n = T_{n-1} + T_{n-2}$. To find the third term which is T_3, plug in $n = 3$. So, $T_3 = T_{3-1} + T_{3-2} \to T_3 = T_2 + T_1$.

Now, substitute the previous terms into the obtained expression.

That is $T_3 = 3 + 1 = 4$. Similarly, by substituting $n = 4$ and $n = 5$ into the given recursive formula, you get:

$$T_4 = T_{4-1} + T_{4-2} \to T_4 = T_3 + T_2 \to T_4 = 4 + 3 = 7,$$
$$T_5 = T_{5-1} + T_{5-2} \to T_5 = T_4 + T_3 \to T_5 = 7 + 4 = 11.$$

The first five terms of the sequence are 1, 3, 4, 7, and 11.

✍ Solve.

1) Find the first five terms of the sequence with the general term
$$a_n = 2a_{n-1} + a_{n-2}; n \geq 3, a_1 = 5 \text{ and } a_2 = -1$$

2) The nth term in a sequence is given by $a_n = 3a_{n-1} + n$ Find the first six terms of this sequence, given that $a_1 = -1$.

3) Find a recursive formula for the sequence $a_n = 1 + 2^n$.

4) Find the 7th term of the following recursive sequence.
$$x_{n+1} = (n-3)x_n + 2; x_2 = 4$$

Evaluate Variable Expressions for Number Sequences

An example of a sequence that has a general term dependent on the index value is an arithmetic sequence. In an arithmetic sequence, the difference between any two consecutive terms is always the same. To find the general term of an arithmetic sequence, we can use the following formula: $x_i = a + (i-1)d$, where a is the first term of the sequence, d is the common difference, and i is the index of the term. By plugging in different values for i, we can find the specific values of the terms in the sequence.

For example, given the sequence 2, 6, 10, 14, 18, $\cdots$, we can see that the common difference is 4. Using the formula above, we can find that the general term of this sequence is $x_i = 2 + (i-1)4$. By plugging in different values for i, such as $i = 1, 2, 3, 4, 5$, and 6 we can find that the first term is 2, the second term is 6, the third term is 10, the fourth term is 14, and the fifth term is 18, respectively. Similarly, we can also find the general term of a geometric sequence. In a geometric sequence, the ratio of any two consecutive terms is always the same. To find the general term of a geometric sequence, we can use the formula $x_i = ar^{i-1}$, where a is the first term of the sequence, r is the common ratio, and i is the index of the term. By plugging in different values for i, we can find the specific values of the terms in the sequence.

In this way we can easily find the terms of the sequence by plugging in different values for the index value in the general term, making it easy to understand and work with sequences.

Let's review an example:

Find the first three terms of the sequence with the general term $a_n = (-1)^{n-1}n$, where n represents the position of a term in the sequence and $n \geq 1$.

Since $n \geq 1$, the sequence starts at one. So, to find the first term, plug-in $n = 1$. Then: $a_1 = (-1)^{1-1} \times 1 = 1$.

In a similar way, enter the natural numbers in order to find the other terms of the sequence. Therefore,

Plug in $n = 2$ as $a_2 = (-1)^{2-1} \times 2 = -2$.

Plug in $n = 3$ as $a_3 = (-1)^{3-1} \times 3 = 3$.

The first three terms of the sequence are $1, -2,$ and 3.

Find the first five terms of the sequence with the general term $x_k = \left(-\frac{1}{2}\right)^k$, where k represents the position of a term in the sequence and starts with $k = 1$.

To find the first term, plug-in $k = 1$. Then: $x_1 = \left(-\frac{1}{2}\right)^1 = -\frac{1}{2}$.

In the same way, to find the 2nd term, plug-in $k = 2$. So, $x_2 = \left(-\frac{1}{2}\right)^2 = \frac{1}{4}$.

To find the 3rd term, plug-in $k = 3 \to x_3 = \left(-\frac{1}{2}\right)^3 = -\frac{1}{8}$.

Finally, $k = 4 \to x_3 = \left(-\frac{1}{2}\right)^4 = \frac{1}{16}$, and $k = 5 \to x_3 = \left(-\frac{1}{2}\right)^5 = -\frac{1}{32}$.

The first five terms of the sequence are $-\frac{1}{2}, \frac{1}{4}, -\frac{1}{8}, \frac{1}{16},$ and $-\frac{1}{32}$.

✍ Given two terms in an arithmetic sequence find the term named in the problem.

5) $a_{12} = 75$ and $a_7 = 40$
   Find a_{10}

6) $a_{10} = 36$ and $a_{20} = 156$
   Find a_{16}

✍ Given two terms in a geometric sequence find the term named.

7) $a_3 = -24$ and $a_7 = -384$
   Find a_5

8) $a_5 = 2$ and $a_7 = \frac{1}{2}$
   Find a_2

✍ Given the explicit formula. Find the first four terms and the 7th term.

9) $a_n = 2 + 3n$

10) $a_n = 2 \cdot a_{n-1}; a_1 = -3$

11) $a_n = (-3)^n + 1$

12) $a_n = a_{n-1} \cdot a_{n-2}; a_1 = -1, a_2 = 2$

13) $a_n = 81 \cdot \left(\frac{1}{3}\right)^{n+1}$

bit.ly/3ZV4MSr

Write Variable Expressions for Arithmetic Sequences

A sequence where the difference between consecutive terms is constant is known as an arithmetic sequence. To write the variable expression for an arithmetic sequence, the following steps should be taken:

Step 1: Identify the initial value, such as $x_1 = a$. This initial value is the first term of the sequence and is usually given in the problem, but if it is not, you will need to find it by observing the first term of the sequence.

Step 2: Determine the common difference, represented as d, between the terms. The common difference is the value that is added or subtracted to each term to get the next term in the sequence.

For example, if the sequence is 2, 5, 8, and 11, the common difference is 3, since each term is increased by 3 to get the next term.

Step 3: Add the first term to the product of the common difference multiplied by the position of the term in the sequence. The position of the term in the sequence is represented by the index of the term, starting from 1.

For example, the first term has an index of 1, the second term has an index of 2, and so on.

So, the variable expression for the nth term of an arithmetic sequence is:

$$x_n = a + (n - 1)d.$$

It's important to note that this method will only work if the sequence is arithmetic in nature, if not the sequence is not arithmetic.

For example, if the sequence is 2, 4, 6, 8, 10, 12, $\cdots$, the initial value $a = 2$, and the common difference $d = 2$, and the variable expression for the nth term of the sequence is $x_n = 2 + (n - 1)2 = 2n$.

Let's review an example:

Write the variable expression for the following sequence 3, 6, 9, 12, 15, $\cdots$.

According to the sequence, the common difference between the consecutive terms is $d = 3$. By dividing the terms by 3, obtains the sequence as 1, 2, 3, 4, 5, ⋯. Actually, the variable expression of this sequence is $a_n = 3n$. Since the first term is $a_1 = 3$, so, the expression $a_n = 3n$ is defined for $n \geq 1$.

Write the general formula for the following sequence 5, 6, 7, 8, 9, ⋯ in terms of variable k.

See that, the common difference between the consecutive terms is $d = 1$.

Subtract 4 from the terms of the sequence, then the obtained the sequence as:

1, 2, 3, 4, 5, ⋯.

Therefore, the general formula of this sequence in terms of variable k is $a_k = 4 + k$ such that $k \geq 1$.

Find the variable expression corresponding to the following sequence.

$$3, 7, 11, 15, 19, \cdots$$

Let $x_1, x_2, x_3, \cdots$ equivalent to the sequence of the content question. So, the first term is $x_1 = 3$. Look for the difference between consecutive terms.

Since the difference of the consecutive terms is the same value equal to $d = 4$. The sequence is arithmetic and it can be rewritten as follows:

$3, 3 + 4, 3 + 2 \times 4, 3 + 3 \times 4, 3 + 4 \times 4, \cdots$.

It seems that the general rule is $x_n = 3 + 4n$. Since the first term in the sequence is 3, the term x_n is not defined for $n \leq 1$.

✎ Determine if the sequence is arithmetic. If it is, find the common difference.

14) 11, 15, 19, 23, ⋯

15) −9, −7, −5, −3, ⋯

16) 19, 14, 10, 7, ⋯

17) 5, 8, 11, 14, ⋯

✎ Solve.

18) An arithmetic sequence has an 8th term of 25 and an 11th term of 34. Find the common difference.

Write Variable Expressions for Geometric Sequences

A sequence where the ratio of consecutive terms is constant is known as a geometric sequence. To write the variable expression for a geometric sequence, the following steps should be taken:

Step 1: Identify the initial value, such as $x_1 = a$. This initial value is the first term of the sequence and is usually given in the problem, but if it is not, you will need to find it by observing the first term of the sequence.

Step 2: Determine the common ratio, represented as r, for the consecutive terms. The common ratio is the value that the terms are multiplied by to get the next term in the sequence.

For example, if the sequence is 2, 4, 8, and 16, the common ratio is 2, since each term is multiplied by 2 to get the next term.

Step 3: Multiply the first term by the product of the common ratio raised to the power of the position of the term in the sequence. The position of the term in the sequence is represented by the index of the term, starting from 1.

For example, the first term has an index of 1, the second term has an index of 2, and so on. So, the variable expression for the nth term of a geometric sequence is $x_n = ar^{n-1}$.

It's important to note that this method will only work if the sequence is geometric in nature, if not the sequence is not geometric.

Let's review some examples:

Write an equation to describe the following sequence 2, 4, 8, 16, 32, $\cdots$.

The given sequence is a geometric sequence since the ratio of consecutive terms is constant. To write the equation that describes this sequence, we can use the variable expression for the nth term of a geometric sequence: $x_n = ar^{n-1}$.

Here: $x_1 = 2$, $a = 2$, and the common ratio $r = 2$, since each term is multiplied by 2 to get the next term. So, the equation that describes this sequence is

$$x_n = 2(2)^{n-1} = 2^n.$$

So, the equation that describes the given sequence is $x_n = 2^n$.

In summary:

- The given recursive sequence $x_1, x_2, x_3, \cdots$ such that the ratio of the consecutive terms is equal to the same value, is called a geometric sequence.
- To write the variable expression for the geometric sequences, follow these steps:
    - Specify the initial values as $x_1 = a$.
    - Determine $r =$ the common ratio of the consecutive terms.
    - Multiply the first term by the product of the number of terms in the common ratio.

Write the general formula for the following sequence $6, 18, 54, 162, 486, \cdots$ in terms of variable i.

By dividing the consecutive terms by previous terms, you notice that the common ratio is $r = 3$. According to the terms of the sequence, you can see that all terms have the common multiple of 2. Then the initial value is $x_1 = 2$. Therefore, by multiplying the first term by the common ratio, we get: $x_i = 2(3)^i$, where $i \geq 1$.

✍ Determine if the sequence is geometric. If it is, find the common ratio.

19) $-1, 6, 8, -36, 216, \cdots$

20) $-2, -2, -4, -8, -16, \cdots$

21) $1, 0.5, 0.25, 0.125, 0.625, \cdots$

22) $9, -3, 1, -\frac{1}{3}, \frac{1}{9}, \cdots$

Write a Formula for a Recursive Sequence

To determine the general formula for a recursive sequence, the following steps should be taken:

Step 1: Identify the initial values, such as a_1 or a_2. These initial values are usually given in the problem, but if they are not, you will need to find them by observing the first few terms of the sequence.

Step 2: Observe the relationship between the previous terms, such as a_{n-1} or a_{n-2}, that is consistent throughout the sequence by evaluating the differences and ratios of consecutive terms.

For example, you might notice that the nth term is always twice the $(n-1)$th term, or that the nth term is always the sum of the previous two terms.

Step 3: Express this relationship as a function of the previous term or terms.

For example, if you notice that the nth term is always twice the $(n-1)$th term, you would express this as a function as: $a_n = 2a_{n-1}$.

Step 4: You can then use this function to express the general formula of the recursive sequence which is $a_n = f(a_{n-1})$ or $a_n = f(a_{n-1}, a_{n-2})$, and so on.

It's important to note that this method will only work if the relationship between the terms is consistent throughout the sequence, if not the sequence is not recursive.

Let's review an example:

Find the recursive formula corresponding to the following sequence.

$$12, 17, 22, 27, 32, \cdots$$

The recursive formula can be found by observing the relationship between consecutive terms in the given sequence.

Notice that the difference between consecutive terms is 5.

So, the recursive formula could be $a_n = a_{n-1} + 5$.

To confirm that this is the correct recursive formula, we can substitute it back into the sequence:

$$a_1 = 12, a_2 = a_1 + 5 = 12 + 5 = 17, a_3 = a_2 + 5 = 17 + 5 = 22,$$
$$a_4 = a_3 + 5 = 22 + 5 = 27, a_5 = a_4 + 5 = 27 + 5 = 32.$$

We can see that this recursive formula generates the given sequence, so the recursive formula corresponding to the sequence is $a_n = a_{n-1} + 5$.

Find the recursive formula for the following sequence.

$$1, 3, 9, 27, 81, \cdots$$

Here, let the first term of this sequence $a_1 = 1$. Looking at the terms in this sequence, you can see that each term has an equal ratio to the previous term. Thus, the following relationship is obtained: $\frac{a_n}{a_{n-1}} = 3$. Now, rewrite this rule in terms of a_{n-1}. Therefore, $a_n = 3a_{n-1}$. You know that $a_1 = 1$. It means that the terms a_{n-1} define for $n \geq 2$.

Find the recursive formula corresponding to the following sequence.

$$1, 3, 2, -1, -3, -4, \cdots$$

Here, after evaluating the terms in the sequence, since the differences and ratios of consecutive terms are not the same value, therefore, look for another relationship between the previous term or even terms of the sequence. However, by evaluating the difference between the terms of the sequence, you can see that the difference of each term from the previous term makes the following sequence: $2, -1, -3, -4, \cdots$. Actually, the term of the last sequence has the formula as $a_n - a_{n-1}$. Therefore, the obtained formula is the recursive formula of the content of question $a_{n+1} = a_n - a_{n-1}$, where $n \geq 2$ $a_1 = 1$, and $a_2 = 3$.

✎ Solve.

23) Write a recursive formula for the following sequence that is valid for $n \geq 2$: $1, 3, 5, 11, 21, 43, \cdots$.

Chapter 13: Answers

1) $5, -1, 3, 5, 13$

2) $-1, -1, 0, 4, 17, 57$

3) $x_n = 2x_{n-1} - 1; x_1 = 3$

4) 32

5) 61

6) 108

7) -96

8) 16

9) $5, 8, 11, 14$ and $x_7 = 23$

10) $-3, -6, -12, -24$ and $x_7 = -192$

11) $-2, 10, -26, 82$ and $x_7 = -2,186$

12) $-1, 2, -2, -4$ and $x_7 = -256$

13) $9, 3, 1, \frac{1}{3}$ and $x_7 = \frac{1}{81}$

14) It's an arithmetic sequence with difference $d = 4$.

15) It's an arithmetic sequence with common $d = 2$.

16) It's not an arithmetic sequence.

17) It's an arithmetic sequence with common $d = 3$.

18) $d = 3$

19) It's not a geometric sequence.

20) It's not a geometric sequence.

21) It's a geometric sequence with common ratio $r = 0.5$.

22) It's a geometric sequence with common ratio $r = -\frac{1}{3}$.

23) $a_n = a_{n-1} + 2a_{n-2}; a_1 = 1, a_2 = 3$ and $n \geq 3$

Time to Test

Time to refine your High School algebra skills with a practice test

Take a High School Algebra I test to simulate the test day experience. After you've finished, score your test using the answer keys.

Before You Start

- You'll need a pencil and a calculator to take the test.
- For multiple questions, there are five possible answers. Choose which one is best.
- It's okay to guess. There is no penalty for wrong answers.
- Use the answer sheet provided to record your answers.
- **Scientific calculator is permitted for High School Algebra I Test.**
- After you've finished the test, review the answer key to see where you went wrong.

Good Luck

High School Algebra I Practice Test 1

2024

Total number of questions: 60

Total time: No time limit

Calculator is permitted for High School Algebra I Test.

High School Algebra I Practice Test Answer Sheet
Remove (or photocopy) this answer sheet and use it to complete the practice test.

High School Algebra I Practice Test 1 Answer Sheet

#		#		#	
1	Ⓐ Ⓑ Ⓒ Ⓓ Ⓔ	21		41	Ⓐ Ⓑ Ⓒ Ⓓ Ⓔ
2		22	Ⓐ Ⓑ Ⓒ Ⓓ Ⓔ	42	Ⓐ Ⓑ Ⓒ Ⓓ Ⓔ
3	Ⓐ Ⓑ Ⓒ Ⓓ Ⓔ	23	Ⓐ Ⓑ Ⓒ Ⓓ Ⓔ	43	Ⓐ Ⓑ Ⓒ Ⓓ Ⓔ
4	Ⓐ Ⓑ Ⓒ Ⓓ Ⓔ	24		44	Ⓐ Ⓑ Ⓒ Ⓓ Ⓔ
5	Ⓐ Ⓑ Ⓒ Ⓓ Ⓔ	25	Ⓐ Ⓑ Ⓒ Ⓓ Ⓔ	45	Ⓐ Ⓑ Ⓒ Ⓓ Ⓔ
6	Ⓐ Ⓑ Ⓒ Ⓓ Ⓔ	26	Ⓐ Ⓑ Ⓒ Ⓓ Ⓔ	46	Ⓐ Ⓑ Ⓒ Ⓓ Ⓔ
7	Ⓐ Ⓑ Ⓒ Ⓓ Ⓔ	27	Ⓐ Ⓑ Ⓒ Ⓓ Ⓔ	47	
8	Ⓐ Ⓑ Ⓒ Ⓓ Ⓔ	28	Ⓐ Ⓑ Ⓒ Ⓓ Ⓔ	48	Ⓐ Ⓑ Ⓒ Ⓓ Ⓔ
9	Ⓐ Ⓑ Ⓒ Ⓓ Ⓔ	29	Ⓐ Ⓑ Ⓒ Ⓓ Ⓔ	49	Ⓐ Ⓑ Ⓒ Ⓓ Ⓔ
10	Ⓐ Ⓑ Ⓒ Ⓓ Ⓔ	30	Ⓐ Ⓑ Ⓒ Ⓓ Ⓔ	50	Ⓐ Ⓑ Ⓒ Ⓓ Ⓔ
11	Ⓐ Ⓑ Ⓒ Ⓓ Ⓔ	31	Ⓐ Ⓑ Ⓒ Ⓓ Ⓔ	51	Ⓐ Ⓑ Ⓒ Ⓓ Ⓔ
12	Ⓐ Ⓑ Ⓒ Ⓓ Ⓔ	32		52	Ⓐ Ⓑ Ⓒ Ⓓ Ⓔ
13	Ⓐ Ⓑ Ⓒ Ⓓ Ⓔ	33	Ⓐ Ⓑ Ⓒ Ⓓ Ⓔ	53	Ⓐ Ⓑ Ⓒ Ⓓ Ⓔ
14	Ⓐ Ⓑ Ⓒ Ⓓ Ⓔ	34	Ⓐ Ⓑ Ⓒ Ⓓ Ⓔ	54	Ⓐ Ⓑ Ⓒ Ⓓ Ⓔ
15	Ⓐ Ⓑ Ⓒ Ⓓ Ⓔ	35	Ⓐ Ⓑ Ⓒ Ⓓ Ⓔ	55	Ⓐ Ⓑ Ⓒ Ⓓ Ⓔ
16		36	Ⓐ Ⓑ Ⓒ Ⓓ Ⓔ	56	Ⓐ Ⓑ Ⓒ Ⓓ Ⓔ
17	Ⓐ Ⓑ Ⓒ Ⓓ Ⓔ	37	Ⓐ Ⓑ Ⓒ Ⓓ Ⓔ	57	Ⓐ Ⓑ Ⓒ Ⓓ Ⓔ
18	Ⓐ Ⓑ Ⓒ Ⓓ Ⓔ	38	Ⓐ Ⓑ Ⓒ Ⓓ Ⓔ	58	Ⓐ Ⓑ Ⓒ Ⓓ Ⓔ
19	Ⓐ Ⓑ Ⓒ Ⓓ Ⓔ	39	Ⓐ Ⓑ Ⓒ Ⓓ Ⓔ	59	Ⓐ Ⓑ Ⓒ Ⓓ Ⓔ
20	Ⓐ Ⓑ Ⓒ Ⓓ Ⓔ	40	Ⓐ Ⓑ Ⓒ Ⓓ Ⓔ	60	Ⓐ Ⓑ Ⓒ Ⓓ Ⓔ

1) If $a = 2x + 6$ and $b = x - 2$, which of the following expresses a in terms b?

   A. $\frac{b+3}{2}$

   B. $b + 3$

   C. $2b - 2$

   D. $2b + 10$

   E. $\frac{b+2}{3}$

2) $f(x) = ax^2 + bx + c$ is a quadratic function where a, b, and c are constant. The value of x of the point of intersection of this quadratic function and linear function $g(x) = 2x - 3$ is 2. The vertex of $f(x)$ is at $(-2, 5)$. What is the product of a, b, and c?

3) The first six terms in a sequence are: 7, 11, 15, 19, 23, 27, ⋯. If the pattern continues, which expression can be used to find the nth term in the sequence?

   A. $4n + 3$

   B. $4n + 7$

   C. $4n + 11$

   D. $4n + 15$

   E. $4n + 13$

4) The table represents some points on the graph of an exponential function.

x	y
-1	16
0	24
1	36
2	54
3	81

Which function represents this relationship?

A. $f(x) = 24\left(\frac{3}{2}\right)^x$

B. $f(x) = 32\left(\frac{3}{2}\right)^x$

C. $f(x) = \frac{3}{2}\left(\frac{2}{3}\right)^x$

D. $f(x) = 32\left(\frac{2}{3}\right)^x$

E. $f(x) = 24\left(\frac{2}{3}\right)^x$

5) The graph of the exponential function h in the xy−plane, where $y = h(x)$, has a y−intercept of d, where d is a positive constant. Which of the following could define the function h?

A. $h(x) = -3(d)^2$
B. $h(x) = 3(x)d$
C. $h(x) = d(-x)^3$
D. $h(x) = d(3x)^2$
E. $h(x) = d(3)^x$

6) An influencer has two different jobs. His combined work schedules consist of less than 65 hours per week.

Which graph best represents the solution set for all possible combinations of x, the number of hours he worked at his first job, and y, the number of hours he worked at his second job, in one week?

A.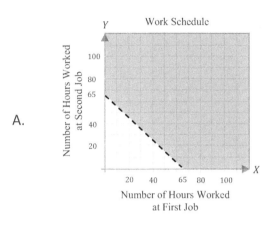

B.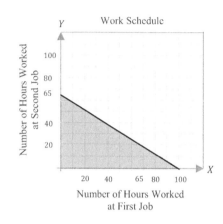

C.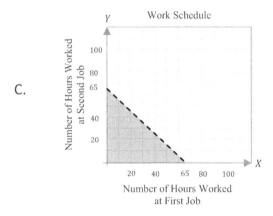

D.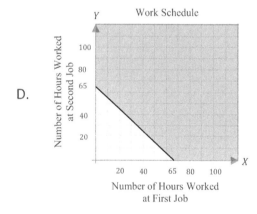

E.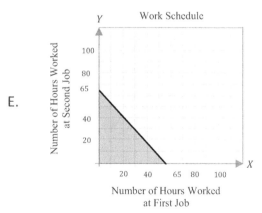

7) Which graph represents the inverse of $y = -x$?

A.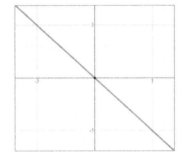

B.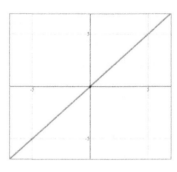

C.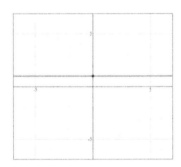

D.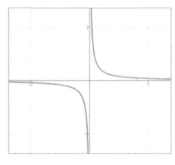

E.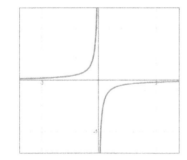

8) If $|a| < 1$, then which of the following is true? ($b > 0$)?

   I. $-b < ba < b$

   II. $-a < a^2 < a$ if $a < 0$

   III. $-5 < 2a - 3 < -1$

A. I only

B. II only

C. I and III only

D. III only

E. I, II and III

9) Which of the following is equal to $m^{\frac{1}{2}}n^{-2}m^4n^{\frac{2}{3}}$?

A. $m^2 n^{-\frac{4}{3}}$

B. $\dfrac{1}{m^{\frac{9}{4}}n^{\frac{4}{3}}}$

C. $\dfrac{m^{\frac{9}{4}}}{n^{\frac{4}{3}}}$

D. $\dfrac{m^{\frac{9}{2}}}{n^{\frac{4}{3}}}$

E. $m^{-\frac{4}{3}}n^2$

10) Divide: $\dfrac{16n^6 - 32n^2 + 8n}{8n}$.

A. $2n^5 - 4n$

B. $2n^6 - 4n^2$

C. $2n^5 - 4n + 1$

D. $2n^6 - 4n^2 + 1$

E. $2n^6 - 4$

11) If a is an integer and 2.46×10^a is a number between 2,000 and 3,000, what is the value of a?

A. 1

B. 2

C. 3

D. 4

E. 5

12) If $f(x) = 3^x$ and $g(x) = \log_3 x$, which of the following expressions is equal to $f(3g(p))$?

A. $3P$

B. $\dfrac{p^3}{3}$

C. p^3

D. p^9

E. $\dfrac{p}{3}$

13) Which of the following is equal to $b^{\frac{3}{5}}$?

A. $b^{\frac{5}{3}}$

B. $\sqrt{b^{\frac{5}{3}}}$

C. $\sqrt{b^{\frac{3}{5}}}$

D. $\sqrt[3]{b^5}$

E. $\sqrt[5]{b^3}$

14) Which graph best represents $f(x) = 0.2(25)^x$?

A.

B.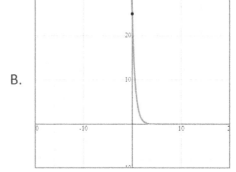

C.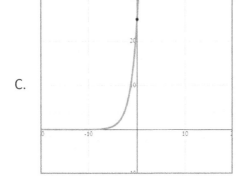

D.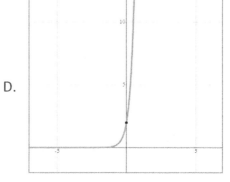

E.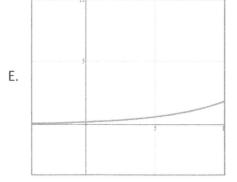

15) Solve this equation for x: $e^{2x} = 12$.

   A. $\frac{1}{2} ln(12)$

   B. $ln(12)$

   C. $ln(2)$

   D. $\frac{1}{2} ln(2)$

   E. $2\, ln(2)$

16) What is the negative solution to the equation $0 = \frac{2x^2}{5} - 10$?

17) Train passengers pay $150 to travel to a certain destination. For this price, customers may bring 3 bags on board. There is a fee of $20 for each additional bag a passenger wants to bring. Which function can be used to find the amount in dollars a passenger has to pay to travel with b bags, where $b \geq 3$?

   A. $c = 20b + 150$

   B. $c = 20(b - 3) + 150$

   C. $c = \left(\frac{1}{20}\right) b + 150$

   D. $c = \left(\frac{1}{2}\right)(b - 3) + 150$

   E. $c = \left(\frac{1}{20}\right)(b - 3) + 150$

18) Find the axis of symmetry of the function $g(x) = -\frac{1}{8}(x - 1)^2 - 3$.

   A. $y = 3$

   B. There is no axis of symmetric.

   C. $x = 3$

   D. $y = -3x + 1$

   E. $x = 1$

19) One of the zeros of the function $f(x) = x^3 - 9x^2 + 20x$ is zero. What is one of the other zeros of this function?

A. -4 and -5

B. -5

C. $0, 4, -5$

D. 4

E. 4 and 5

20) Quadratic functions f and h are shown below.
$$f(x) = 3x^2 - 6$$
$$h(x) = 3x^2 + c$$

For what value of c will the graph of h be 12 units above the graph of f?

A. -12

B. -6

C. 0

D. 6

E. 12

21) If $x = 9$, what is the value of y in the following equation? $2y = \frac{2x^2}{3} + 6$

22) Create a cubic function that has roots at $x = -1, 1$, and 3.

A. $f(x) = x^3 + 3x^2 + x + 3$

B. $f(x) = x^3 - 3x^2 - x - 3$

C. $f(x) = x^3 - 3x^2 - x + 3$

D. $f(x) = x^3 - 6x^2 - 2x - 3$

E. $f(x) = x^3 - 6x^2 - 2x + 3$

23) The perimeter of the trapezoid below is 36 cm. What is its area?
    A. 24 cm^2
    B. 36 cm^2
    C. 48 cm^2
    A. 57 cm^2
    E. 70 cm^2

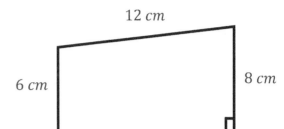

24) What is the value of x in the following equation?
$$\frac{x^2-9}{x+3} + 2(x+4) = 15$$

25) If x represents a negative number, which of the following must be true?
    A. $2x^2 < 0$
    B. $\frac{x^2}{8} < 0$
    C. $x^2 > x^3$
    D. $\frac{x}{8} > 0$
    E. None

26) Which function is best represented by this graph?
    A. $ln(x)$
    B. $ln(x+1)$
    C. $ln(x) - 1$
    D. $ln(x) + 1$
    E. $ln(x-1)$

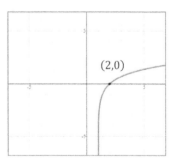

27) The initial value of a car is $35,000. The value of the car will decrease at a rate of 20% each year.

Which graph best models this situation?

A.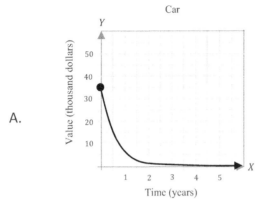

B.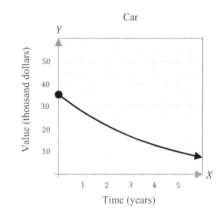

C.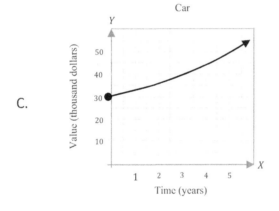

D.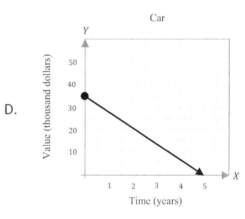

E.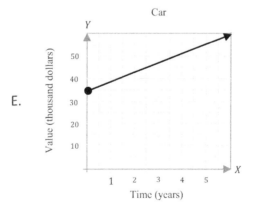

28) The graph of a quadratic function is shown on the grid.

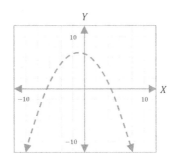

Which equation best represents the axis of symmetry?

A. $x = -1$
B. $y = -1$
C. $x = 0$
D. $y = 0$
E. $y = 7$

29) If the function is defined as $f(x) = bx^2 + 15$, b is a constant, and $f(2) = 35$. What is the value of $f(3)$?

A. 25
B. 35
C. 45
D. 60
E. 65

30) How many x−intercepts does the graph of $y = \frac{x-1}{1-x^2}$ have?

A. 0
B. 1
C. 2
D. 3
E. 4

31) The late fee for overdue books at a library is $0.2 per day per book, with a maximum late fee of $4.00 per book. Which graph models the total late fee for 5 books that were checked out on the same day and are overdue?

A.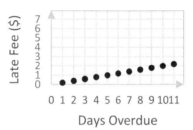

B.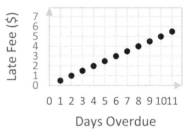

C.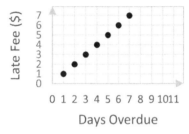

D.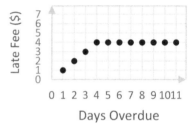

E.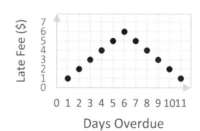

32) Quadratic function k can be used to show the height in meters of a rocket from the ground t seconds after it was launched. The graph of the function is shown.

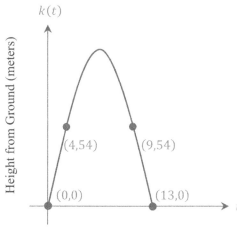

What is the maximum value of the graph of the function?

☐

33) If the function f is defined by $f(x) = x^2 + 2x - 10$, which of the following is equivalent to $f(4t^2)$?

A. $3t^4 + 6t^2 - 10$
B. $3t^4 + 3t^2 - 10$
C. $3t^4 + 6t^2 + 10$
D. $16t^4 + 8t^2 + 10$
E. $16t^4 + 8t^2 - 10$

34) Which statement best describes these two functions?

$$f(x) = 2x^2 - x + 3$$
$$g(x) = -x^2 + 2x + 1$$

A. The maximum of $f(x)$ is less than the minimum of $g(x)$.
B. The minimum of $f(x)$ is less than the maximum of $g(x)$.
C. The minimum of $f(x)$ is equal to the maximum of $g(x)$.
D. The maximum of $f(x)$ is greater than the minimum of $g(x)$.
E. The minimum of $f(x)$ is greater than the maximum of $g(x)$.

35) If $y = (-3x^3)^2$, which of the following expressions is equal to y?

A. $-6x^5$

B. $-6x^6$

C. $6x^5$

D. $9x^5$

E. $9x^6$

36) A quadratic function is graphed on the grid.

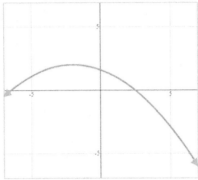

Which answer choice best represents the domain and range of the function?

A. Domain: All real numbers

   Range: $y \geq 2$

B. Domain: $-7 \leq x \leq 7$

   Range: $y < 2$

C. Domain: $y < 2$

   Range: All real numbers

D. Domain: $-7 < x < 7$

   Range: $y \leq 2$

E. Domain: All real numbers

   Range: $y \leq 2$

37) Which of the following is equivalent to $2\sqrt{32} + 2\sqrt{2}$?

A. $\sqrt{2}$

B. 2

C. 4

D. $4\sqrt{2}$

E. $10\sqrt{2}$

38) The function $g(x)$ is defined by a polynomial. Some values of x and $g(x)$ are shown in the table below. Which of the following must be a factor of $g(x)$?

A. x

B. $x - 1$

C. $x + 1$

D. $x - 2$

E. $x + 2$

x	$g(x)$
0	5
1	4
2	0

39) A company collected data for the number of text messages sent and received using a text-message application since March 2018. The table shows the number of text messages sent and received in billions over time. The data can be modeled by a quadratic function.

Number of Followers on the Page, x	Number of Feedbacks of each Post, y
5	9.6
10	12.25
15	16.6
20	22.75
25	30.65
30	40.25

Which function best models the data?

A. $y = -0.025t^2 + 5.02$

B. $y = -0.025t^2 + 0.7t + 7.02$

C. $y = 0.035t^2 + 8.75$

D. $y = 0.035t^2 - 0.03t + 8.75$

E. $y = -0.035t^2 + 8.75$

40) If a and b are two values of p that satisfy the equation $p^2 - 6p + 8 = 0$, with $a > b$, what is the value of $a - b$?

   A. 10

   B. 8

   C. 6

   D. 4

   E. 2

41) Solve the equation: $log_3(x + 20) - log_3(x + 2) = 1$.

   A. 3

   B. 7

   C. 14

   D. 22

   E. 31

42) Simplify. $7x^2y^3(2x^2y)^3 =$

   A. $14x^8y^4$

   B. $14x^8y^6$

   C. $56x^4y^8$

   D. $56x^4y^6$

   E. $56x^8y^6$

43) The figure below shows the graph of function f. What is the value of $f(f(2))$?

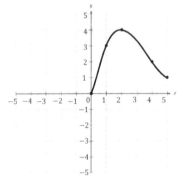

   A. 0

   B. 1

   C. 2

   D. 3

   E. 5

44) Which expression is equivalent to a factor of $3n^2 + 5n - (n+3)(3n-5)$?

A. $3n - 5$

B. $n + 3$

C. $n - 15$

D. $n + 5$

E. $n + 15$

45) Two functions are given below.

$$g(x) = \frac{9}{11}x + 1$$
$$h(x) = \frac{3}{2}x + 1$$

How does the graph of g compare with the graph of h?

A. The graph of g has a different y-intercept than the graph of h.

B. The graph of g is less steep than the graph of h.

C. The graph of g is steeper than the graph of h.

D. The graph of g is parallel to the graph of h.

E. The graph of g is perpendicular to the graph of h.

46) Which statement about the absolute value parent function is Not true?

A. Its graph is symmetrical about the x-axis.

B. Its graph is symmetrical about the y-axis.

C. Its domain is the set of all real numbers.

D. Its range is the set of all non-negative numbers.

E. Its graph is non-negative.

47) The expression $6x^2 + 4x - 10$ can be written in factored form as $(2x - m)(3x + 5)$, where m represents a number. What is the value of m?

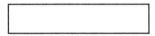

48) Which statement best describes these two functions?

$$f(x) = 2x^2 - 4x$$
$$g(x) = (x-1)^2 - 1$$

A. They have no common points.

B. They have the same x-intercepts.

C. The maximum of $f(x)$ is the same as the minimum of $g(x)$.

D. They have the same minimum.

E. They have one common point.

49) If $(ax + 4)(bx + 3) = 10x^2 + cx + 12$ for all values of x and $a + b = 7$, what are the two possible values for c?

   A. 20, 22
   B. 22, 21
   C. 23, 26
   D. 26, 24
   E. 24, 23

$$4x^2 + 6x - 3, 3x^2 - 5x + 8$$

50) Which of the following is the sum of the two polynomials shown above?

   A. $x^2 + 5x + 4$
   B. $4x^2 - 6x + 3$
   C. $4x^2 + 5$
   D. $5x^2 + 3x + 4$
   E. $7x^2 + x + 5$

51) The graph of $h(x) = 3^x$ was transformed to create the graph of $k(x) = -4(3^x)$. Which of the following describes the transformation from $h(x)$ to $k(x)$?

   A. A reflection over the y −axis and a vertical stretch
   B. A reflection over the x −axis and a vertical stretch
   C. A reflection over the y −axis and a horizontal stretch
   D. A reflection over the x −axis and a horizontal stretch
   E. A reflection over the x −axis and a horizontal stretch

52) Given $f(x^3) = 2x - 5$, for all values of x, what is the value of $f(8)$?

   A. -11
   B. -1
   C. 1
   D. 9
   E. 11

53) The graph of a quadratic function is shown on the grid.

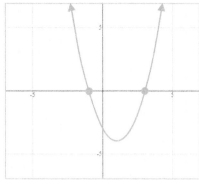

Which function is best represented by this graph?

A. $y = x^2 + 2x - 3$

B. $y = x^2 - 2x + 3$

C. $y = x^2 - 2x - 3$

D. $y = -x^2 + 2x + 3$

E. $y = -x^2 - 3x + 2$

54) Which expression is a factor of $15x^2 - 12x - 3$?

A. $x + 1$

B. $5x + 1$

C. $5x - 1$

D. $3x - 1$

E. $3x + 1$

55) Which table shows y as a function of x?

A.
x	-2	-2	-2	-2
y	0	4	15	-4

B.
x	-7	1	1	12
y	1	3	10	4

C.
x	-12	-1	0	2
y	2	2	2	2

D.
x	-17	0	15	-17
y	0	4	15	-4

E.
x	-2	0	1	0
y	4	-2	11	0

56) Which expression is equivalent to $35r^2 - 28r$?

 A. $7r(5r - 4)$

 B. $7r(5r + 4)$

 C. $7(5r + 4)$

 D. $7(5r - 4)$

 E. $5r(7r - 4)$

57) Which expression is a factor of $25x^2 + 5x + 1$?

 A. $5x - 1$

 B. $5x + 1$

 C. $x - 1$

 D. $25x + 1$

 E. There are no real factors.

58) An exponential function is graphed on the grid.

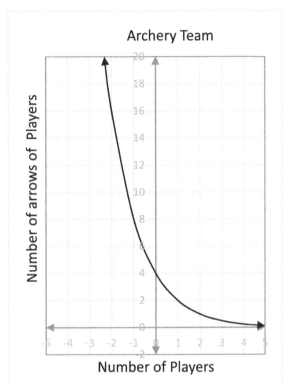

Which function is best represented by the graph?

A. $p(x) = 4(0.25)^x$
B. $p(x) = 4(0.5)^x$
C. $p(x) = (0.75)^x$
D. $p(x) = 2(1.25)^x$
E. $p(x) = 2(2.25)^x$

59) For $i = \sqrt{-1}$, which of the following is equivalent to $\frac{2+3i}{5-2i}$?

A. $\frac{3+2i}{5}$
B. $\frac{4+19i}{29}$
C. $\frac{4+19i}{20}$
D. $5 - 3i$
E. $5 + 3i$

60) If the equation $y = (x+3)(x-9)$ is graphed in the xy-plane, what is the y-coordinate of the parabola's vertex?

A. -36
B. -9
C. -3
D. 3
E. 24

End of High School Algebra I Practice Test 1

High School Algebra I Practice Test 2

2024

Total number of questions: 60

Total time: No time limit

Calculator is permitted for High School Algebra I Test.

High School Algebra I Practice Test Answer Sheet

Remove (or photocopy) this answer sheet and use it to complete the practice test.

High School Algebra I Practice Test 2 Answer Sheet

1	ABCDE	21	ABCDE	41	ABCDE
2	ABCDE	22	ABCDE	42	
3	ABCDE	23	ABCDE	43	ABCDE
4	ABCDE	24	ABCDE	44	ABCDE
5		25	ABCDE	45	ABCDE
6	ABCDE	26	ABCDE	46	ABCDE
7	ABCDE	27	ABCDE	47	ABCDE
8	ABCDE	28	ABCDE	48	ABCDE
9	ABCDE	29	ABCDE	49	ABCDE
10	ABCDE	30	ABCDE	50	ABCDE
11	ABCDE	31	ABCDE	51	
12	ABCDE	32	ABCDE	52	ABCDE
13	ABCDE	33		53	ABCDE
14	ABCDE	34	ABCDE	54	ABCDE
15	ABCDE	35	ABCDE	55	
16	ABCDE	36	ABCDE	56	ABCDE
17	ABCDE	37	ABCDE	57	ABCDE
18	ABCDE	38	ABCDE	58	ABCDE
19	ABCDE	39	ABCDE	59	ABCDE
20	ABCDE	40		60	

High School Algebra I Practice Test 2

1) Which of the following equations relates y to x for the values in the table below?

   A. $y = \frac{1}{2}x + \frac{3}{5}$

   B. $y = \frac{3}{5}x + 2$

   C. $y = \left(\frac{3}{5}\right)^x - 2$

   D. $y = \left(\frac{1}{2}\right)^x + \frac{2}{5}$

   E. $y = \left(\frac{1}{3}\right)^{\frac{3}{5}x}$

x	y
2	$\frac{8}{5}$
3	$\frac{21}{10}$
4	$\frac{13}{5}$
5	$\frac{31}{10}$

2) The equation $x^2 = 4x - 3$ has how many distinct real solutions?

   A. 0

   B. 1

   C. 2

   D. 3

   E. 4

3) Which table does NOT show y as a function of x?

   A.
x	−1.2	2	1.2	−2	2
y	1.5	−8	15	−4	8

   B.
x	0.7	−1.5	1	−0.2	−2.1
y	−0.8	3.9	10.2	−4.5	−0.2

   C.
x	12	13	14	15	16
y	80	70	60	50	40

   D.
x	$\frac{1}{10}$	$\frac{1}{5}$	$\frac{2}{5}$	1	$\frac{8}{5}$
y	10	4	15	−4	7

   E.
x	−1.25	−0.8	1.5	5.25	19.2
y	−1	0	1	5	19

4) The following graph shows the number of bacteria in a laboratory sample in a few hours. Based on this information, which function best shows the number of bacteria x per hour?

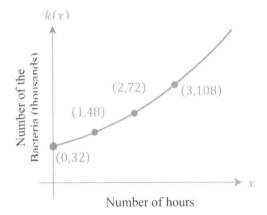

A. $y = (0.48)^x$
B. $y = 48(0.67)^x$
C. $y = 32(1.5)^x$
D. $y = 1.5(32)^x$
E. $y = 0.67(48)^x$

5) $f(x) = 5x^2 + c$ and $g(x) = 4x - 6$, where c is a constant. If $g(f(1)) \times g(2) = 4$, what is the value of c?

6) Which of the following is equal to the expression below?
$$(5x + 2y)(2x - y)$$

A. $4x^2 - 2y^2$
B. $xy - 2y^2$
C. $2x^2 + 6xy - 2y^2$
D. $8x^2 + 2xy - 2y^2$
E. $10x^2 - xy - 2y^2$

7) Two characteristics of quadratic function f are given.
   - The axis of symmetry of the graph of f is $x = 1$.
   - Function f has exactly two zeros.

   Based on this information, which graph could represent f?

A.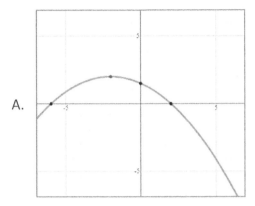

B.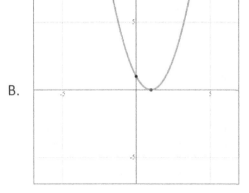

C.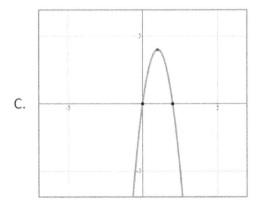

D.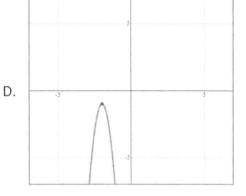

E.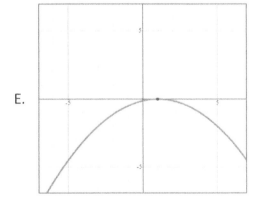

8) Which one is not true for the function $g(x) = 3x^2 - 6x$?
   A. The axis of symmetry of the function g is $x = 1$.
   B. The zeros are 0 and 2.
   C. The factors of g are $3x$ and $x - 2$.
   D. The axis of symmetry of the function g is $y = -3$.
   E. The function has a minimum point with the coordinate $(1, -3)$.

9) Which ordered pair is in the solution set of $y > 3 - \frac{5}{2}x$?
   A. $(0, 3)$
   B. $(0, -6)$
   C. $(6, -4.5)$
   D. $(-1, 0)$
   E. $(0, 0)$

10) What are the zeroes of the function $f(x) = x^3 + 7x^2 + 12x$?
    A. 0
    B. $-4, -3$
    C. $0, 2, 3$
    D. $0, -3, -4$
    E. $0, 3, 4$

$$y < c - x, \; y > x + b$$

11) In the xy-plane, if $(0, 0)$ is a solution to the system of inequalities above, which of the following relationships between c and b must be true?
    A. $c < b$
    B. $c = b$
    C. $c \leq b$
    D. $c = b + c$
    E. $c > b$

12) In 1999, the average worker's income increased by $2,000 per year starting from a $26,000 annual salary. Which equation represents income greater than average?
(I = income, x = number of years after 1999)

A. $I > 2,000x + 26,000$
B. $I > -2,000x + 26,000$
C. $I < -2,000x + 26,000$
D. $I < 2,000x - 26,000$
E. $I < -2,000x - 26,000$

13) Based on the definition of the following function, for which of the following values of x is the function g NOT defined?

$$g(x) = \frac{2x}{x^2 - x - 2}$$

A. 2, 5
B. 3, 2
C. 1, −4
D. −1, 2
E. 3, −4

14) $\frac{5x^2 + 75x - 80}{x^2 - 1}$?

A. $\frac{5x + 75}{x - 1}$
B. $\frac{x + 16}{x + 1}$
C. $\frac{5x + 80}{x + 1}$
D. $\frac{x + 15}{x - 1}$
E. $\frac{x + 1}{5x - 75}$

15) If $x^2 + 6x - r$ is divisible by $(x - 5)$, what is the value of r?

A. 55

B. 56

C. 57

D. 58

E. 59

16) The scatterplot and table show the monthly profit in dollars earned from the sale of a type of home furniture set at nine different prices. The data can be shown by a quadratic function.

Which function best models the data?

x	y
5.7	8.301
5.9	109.15
6	120.4
6.1	130.45
6.5	158.55
6.95	167.06
7.01	166.35
7.8	116.4
8.3	45.83

A. $y = -60.4x^2 + 831.3x - 2693$

B. $y = 0.001x^2 + 35.672$

C. $y = 60.4x^2 - 845.6x - 2876$

D. $y = 0.001x^2 - 0.426x + 35.672$

E. $y = -60.4x^2 + 795.1x - 2448$

17) Which of the following numbers is NOT a solution to the inequality $2x - 5 \geq 3x - 1$?
   A. -2
   B. -4
   C. -5
   D. -8
   E. -12

18) A part of an exponential function is graphed on the grid.

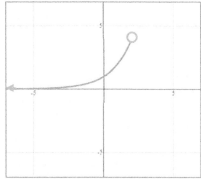

Which inequality best represents the range of the part shown?
   A. $0 \leq y$
   B. $-\infty < x < 2$
   C. $x \leq 2$
   D. $2 \leq x < 4$
   E. $0 < y < 4$

19) If $x^2 + 3$ and $x^2 - 3$ are two factors of the polynomial $12x^4 + n$ and n is a constant, what is the value of n?
   A. -108
   B. -24
   C. 12
   D. 24
   E. 108

20) $(2i - 3) - (2 - i)(i - 2) =$
   A. $2i$
   B. $-2i$
   C. $2i - 1$
   D. $-2i + 1$
   E. $2i + 1$

21) If $f(x^2) = 3x + 4$, for all positive values of x, what is the value of $f(121)$?
   A. -33
   B. -29
   C. 29
   D. 33
   E. 37

22) A basketball player throws a ball toward the basketball net. The graph shows the height in yards of the basketball ball above the ground as a quadratic function of x, the horizontal distance in yards of the basketball ball from the hands of the player.

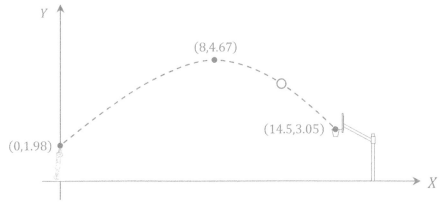

   What is the domain of the function for this solution?
   A. $0 \leq x \leq 4.67$
   B. $0 \leq y \leq 4.67$
   C. $1.98 \leq y \leq 4.67$
   D. $1.98 \leq x \leq 14.5$
   E. $0 \leq x \leq 14.5$

$$f(x) = (5m + 5)x^3 - (m^2 - 16)x$$

23) In the expression above, m is a constant and has a positive value. If the expression is equivalent to nx^3, where n is a constant, what is the value of n?
   A. 4
   B. 12
   C. 16
   D. 18
   E. 25

24) John buys a pepper plant that is 5 inches tall. With regular watering, the plant grows 3 inches a year. Writing John's plant's height as a function of time, what does the y −intercept represent?

A. The y −intercept represents the rate of growth of the plant which is 5 inches.
B. The y −intercept represents the starting height of 5 inches.
C. The y −intercept represents the rate of growth of a plant which is 3 inches per year.
D. There is no y −intercept.
E. There are two y −intercepts.

25) The graph of $y = f(x)$ in the xy −plane is shown below. What is the value of $f(0)$?

A. −2
B. −1
C. 0
D. 1
E. 2

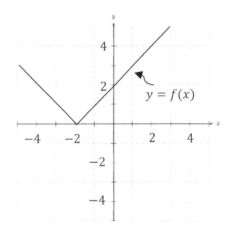

26) The complete graph of function f is shown in the xy −plane below. For what value of x is the value of $f(x)$ at its minimum?

A. −4
B. −3
C. −2
D. 2
E. 3

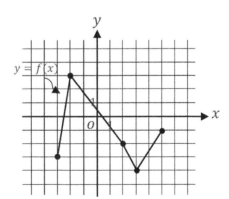

27) The slope and y-intercept of the graph of f were changed to make the graph of g, as shown below.

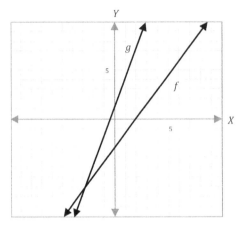

Which statement describes the changes that were made to the graph of f to make the graph of g?

A. The slope was multiplied by -2, and the y-intercept was decreased by 4 to make the graph of g.

B. The slope was multiplied by $-\frac{1}{2}$, and the y-intercept was decreased by 4 to make the graph of g.

C. The slope was multiplied by 2, and the y-intercept was increased by 4 to make the graph of g.

D. The slope was multiplied by $\frac{1}{2}$, and the y-intercept was increased by 4 to make the graph of g.

E. The slope was multiplied by $-\frac{1}{2}$, and the y-intercept was increased by 4 to make the graph of g.

28) What is the solution to the system of equations below?
$$3x + y = 15$$
$$-2x + 4y = 10$$

A. $\left(\frac{25}{7}, \frac{7}{30}\right)$

B. $\left(\frac{25}{7}, \frac{30}{7}\right)$

C. $\left(\frac{25}{7}, -\frac{30}{7}\right)$

D. There is no solution.

E. There is an infinite number of solutions.

29) In the following graph, which of the data points is farthest from the line of best fit (Not shown)?

A. (2,2)
B. (6,1)
C. (5,4)
D. (3,3)
E. No point

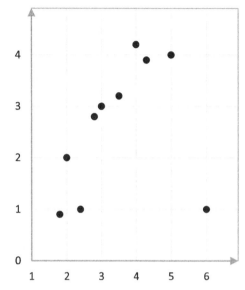

30) In a sequence of numbers, $a_4 = 19$, $a_5 = 23$, $a_6 = 27$, $a_7 = 31$, and $a_8 = 35$. Based on this information, which equation can be used to find the nth term in the sequence, a_n?

A. $a_n = 4n - 3$
B. $a_n = 3n + 4$
C. $a_n = 4n + 3$
D. $a_n = 3n - 4$
E. $a_n = 3n + 3$

31) Which of the following are the solutions for the equation $3x^2 - 4x = 3 - x$?

A. $x = 3$ and $x = 1$
B. $x = 3$ and $x = -\frac{3}{4}$
C. $x = \frac{1-\sqrt{5}}{2}$ and $x = \frac{3-\sqrt{5}}{2}$
D. $x = \frac{3+\sqrt{5}}{2}$ and $x = \frac{3-\sqrt{5}}{2}$
E. $x = \frac{1+\sqrt{5}}{2}$ and $x = \frac{1-\sqrt{5}}{2}$

32) What is the ratio of the minimum value to the maximum value of the following function?
$$f(x) = -3x + 1, -2 \leq x \leq 3$$

A. $-\frac{8}{7}$

B. $-\frac{4}{3}$

C. $\frac{7}{8}$

D. $\frac{8}{7}$

E. $\frac{4}{3}$

33) In the xy-plane, if a point with coordinates (a, b) lies in the solution set of the system of inequalities below, what is the maximum possible value of b?
$$y \leq -15x + 3{,}000$$
$$y \leq 5x$$

34) A conservation agency tracks the stork population by counting the number of nesting sites where the storks lay their eggs. The table shows the number of nesting sites for several years since 2014. The data can be shown by an exponential function.
Which function best models the data?

A. $n(x) = 71{,}235(0.77)^x$

B. $n(x) = 71{,}235(0.89)^x$

C. $n(x) = 35{,}397(1.05)^x$

D. $n(x) = 1.05(35{,}397)^x$

E. $n(x) = 0.89(71{,}235)^x$

Number of Years Since 2014, x	Number of Nesting Sites, $n(x)$
1	55,965
2	43,082
3	38,487
4	29,162
5	22,385
6	19,007
7	16,549

35) What is the area of the following equilateral triangle if the side $AB = 12\ cm$?

A. $8\ cm^2$

B. $6\sqrt{3}\ cm^2$

C. $9\sqrt{3}\ cm^2$

D. $18\sqrt{3}\ cm^2$

E. $36\sqrt{3}\ cm^2$

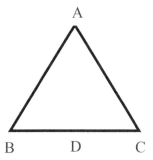

36) If $x^2 - 3x - 10 = 0$ and $x^2 - x - 20 = 0$, what is the value of x?

   A. -4

   B. -2

   C. 2

   D. 4

   E. 5

37) Suppose a type of oak tree has an average height of 45 feet when it is 15 years old. If the tree is more than 15 years old, the average height, h, can be modeled by the function $h = 1.3(a - 15) + 45$, where a is the age of the tree in years. Which statement about this situation is true?

   A. Every additional $1.3\ ft$ of length over $45\ ft$ adds 15 years to the age of this type of oak tree.

   B. For this type of oak tree, the average height increases by $1.3\ ft$ per year throughout its lifetime.

   C. Each additional year of age over 15 years adds $1.3\ ft$ to the average height of this type of oak tree.

   D. For this type of oak tree, the average height increases by $45\ ft$ for every 15 years of growth.

   E. Every additional $14\ ft$ of length over $1.3\ ft$ adds 45 years to the age of this type of oak tree.

38) If a parabola with equation $y = ax^2 + 5x + 10$, where a is constant passes through the point $(2, 12)$, what is the value of a^2?

   A. -4

   B. -2

   C. 1

   D. 2

   E. 4

39) What is the parent graph of the following function and what transformations have taken place on it? $y = -x^2 + 2x$

   A. The parent graph is $y = x^2$, which is reflected across the $x-$axis and is shifted 1 unit up and shifted 1 unit to the right.

   B. The parent graph is $y = x^2$, which is shifted 2 units up.

   C. The parent graph is $y = -x^2$, which is shifted 2 units left.

   D. The parent graph is $y = x^2$, which is reflected across the $x-$axis and is shifted 1 unit to the right and shifted 1 unit to the down.

   E. The parent graph is $y = x^2$, which is shifted 2 units right.

40) What is the value of the y−intercept of the graph $f(x) = 11.5(0.8)^x$?

41) Which expression is equivalent to $0.00035 \times (1.2 \times 10^4)$?
   A. 4.2×10^{-1}
   B. 4.2
   C. 4.2×10
   D. 4.2×10^2
   E. 4200

42) The expression $n^{-3}(n^2)^3$ is equivalent to n^x. What is the value of x?

43) Which expression is equivalent to $k^2 - 17k + 66$?
   A. $(k + 6)(k + 11)$
   B. $(k - 33)(k - 2)$
   C. $(k + 33)(k + 2)$
   D. $(k - 6)(k - 11)$
   E. $(k - 3)(k + 22)$

44) The perimeter of a rectangular garden is 38 meters. The length of the garden can be represented by $(x + 6)$ meters, and its width can be represented by $(2x - 2)$ meters. What are the dimensions of this garden in meters?
   A. Length = 11 meters and width = 8
   B. Length = 11 meters and width = 20
   C. Length = 8 meters and width = 22
   D. Length = 14 meters and width = 16
   E. Length = 8 meters and width = 8

45) Solve: $\frac{3x+6}{x+5} \times \frac{x+5}{x+2} =$.
   A. 1
   B. 2
   C. 3
   D. $\frac{1}{3(x+2)}$
   E. $\frac{x+5}{x+2}$

46) Which graph does not represent y as a function of x?

A.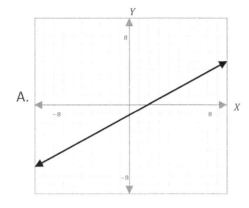

B.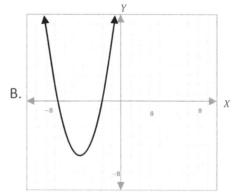

C.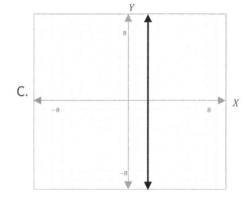

D.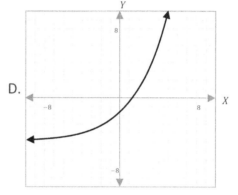

E.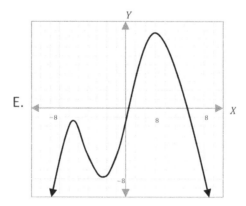

47) The graph of $f(x) = x^2$ is transformed to create the graph of $g(x) = 0.35f(x)$. Which graph best represents f and g?

A.

B.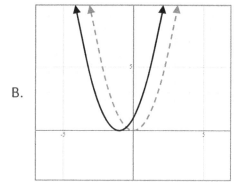

C.

D.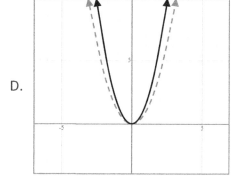

E.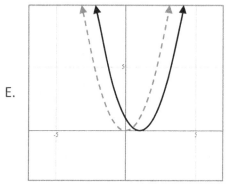

48) The surface area of a cylinder is $150\ \pi\ cm^2$. If its height is $10\ cm$, what is the radius of the cylinder?

   A. $5\ cm$

   B. $10\ cm$

   C. $11\ cm$

   D. $13\ cm$

   E. $15\ cm$

49) Quadratic functions f and g are graphed on the grid. The graph of f was transformed to create the graph of g.

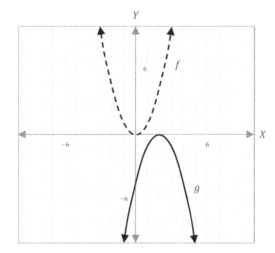

   Which function best represents the graph of g?

   A. $g(x) = (x + 2)^2$

   B. $g(x) = x^2 - 2$

   C. $g(x) = x^2 + 2$

   D. $g(x) = -x^2 - 2$

   E. $g(x) = -(x - 2)^2$

50) Simplify $\frac{6}{\sqrt{12}-3}$.

   A. $\sqrt{12} - 3$

   B. 2

   C. $\sqrt{12} + 3$

   D. $2\sqrt{12}$

   E. $2(\sqrt{12} + 3)$

51) The value of $\frac{8+3x}{7} - \frac{4-4x}{7}$ is how much greater than the value of x?

52) In the diagram below, circle A represents the set of all odd numbers, circle B represents the set of all negative numbers, and circle C represents the set of all multiples of 7. Which number could be replaced with y?

   A. 14
   B. 0
   C. -14
   D. -21
   E. -28

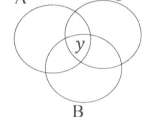

53) Calculate $f(5)$ for the following function f.
$$f(x) = x^2 - 3x$$

   A. 5
   B. 10
   C. 15
   D. 20
   E. 25

54) What is the sum of all values of n that satisfies $2n^2 + 16n + 24 = 0$?

   A. 8
   B. 4
   C. 0
   D. -4
   E. -8

55) What is the positive solution to this equation? $6(x-1)^2 = 41 - x$

56) If $f(x) = \frac{10x-3}{6}$ and $f^{-1}(x)$, is the inverse of $f(x)$, what is the value of $f^{-1}(2)$?

   A. $\frac{6}{17}$
   B. $\frac{3}{2}$
   C. $\frac{2}{3}$
   D. $\frac{17}{6}$
   E. 3

57) If $4x = \frac{48}{3}$, what is the value of 7^{x-2}?

A. 7
B. 14
C. 34
D. 49
E. 343

58) The set of ordered pairs below represents some points on the graph of function f.
$$\{(2,3)(0,-1)(4,7)(-1,-3)(-3,-7)\}$$
What is the parent function of f?

A. $y = x$
B. $y = 2^x$
C. $y = x^2$
D. $y = \sqrt{x}$
E. $y = \ln x$

59) What is the value of the y-intercept of the graph of $g(x) = 25(1.2)^{x+1}$?

A. 1
B. 1.2
C. 20
D. 25
E. 30

60) The graph of quadratic function f is shown on the grid.

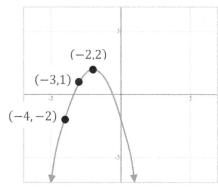

What is the y-intercept of the graph of f?

End of High School Algebra I Practice Test 2

High School Algebra I Practice Tests Answer Keys

Now, it's time to review your results to see where you went wrong and what areas you need to improve.

High School Algebra I Practice Test 1						High School Algebra I Practice Test 2					
1	D	21	30	41	B	1	A	21	E	41	B
2	1	22	C	42	E	2	C	22	E	42	3
3	A	23	E	43	C	3	A	23	E	43	D
4	A	24	$\frac{10}{3}$	44	E	4	C	24	B	44	A
5	E	25	C	45	C	5	-3	25	E	45	C
6	C	26	E	46	A	6	E	26	E	46	C
7	A	27	B	47	2	7	C	27	C	47	A
8	C	28	A	48	B	8	D	28	B	48	A
9	D	29	D	49	C	9	C	29	B	49	E
10	C	30	A	50	E	10	D	30	C	50	E
11	C	31	C	51	B	11	E	31	E	51	$\frac{4}{7}$
12	C	32	63.4	52	B	12	A	32	A	52	D
13	E	33	E	53	C	13	D	33	750	53	B
14	A	34	E	54	B	14	C	34	A	54	E
15	A	35	E	55	C	15	A	35	E	55	$\frac{7}{2}$
16	-5	36	E	56	A	16	A	36	E	56	B
17	B	37	E	57	E	17	A	37	C	57	D
18	E	38	D	58	B	18	E	38	E	58	A
19	E	39	C	59	B	19	A	39	A	59	E
20	D	40	E	60	A	20	B	40	11.5	60	-2

High School Algebra I Practice Tests 1 Explanations

1) **Choice D is correct**

$b = x - 2 \to x = b + 2$

Substitute $b + 2$ for x in the first equation:

$a = 2(b + 2) + 6 \to a = 2b + 4 + 6 = 2b + 10$

2) **The answer is** 1

The intersection of two functions is the point with 2 for x. Then:

$f(2) = g(2)$ and $g(2) = (2 \times (2)) - 3 = 4 - 3 = 1$

Then, $f(2) = 1 \to a(2)^2 + b(2) + c = 1 \to 4a + 2b + c = 1$ (i)

The value of x in the vertex of the parabola is: $x = -\frac{b}{2a} \to -2 = -\frac{b}{2a} \to b = 4a$

(ii)

In the point $(-2, 5)$, the value of the $f(x)$ is 5.

$f(-2) = 5 \to a(-2)^2 + b(-2) + c = 5 \to 4a - 2b + c = 5$ (iii)

Using the first two equations:

$\begin{cases} 4a + 2b + c = 1 \\ 4a - 2b + c = 5 \end{cases} \to$

Equation 1 minus equation 2 is: (i)−(iii) $\to 4b = -4 \to b = -1$ (iv)

Plug in the value of b in the second equation: $b = 4a \to a = \frac{b}{4} = -\frac{1}{4}$

Plug in the values of a and be in the first equation. Then:

$\to 4\left(\frac{-1}{4}\right) + 2(-1) + c = 1 \to -1 - 2 + c = 1 \to c = 1 + 3 \to c = 4$

The product of a, b, and $c = \left(-\frac{1}{4}\right) \times (-1) \times 4 = 1$.

3) **Choice A is correct**

In the given sequence, each term is obtained by adding 4 to the previous term. Therefore, to find the nth term in the sequence, we can use the formula:

$a_n = a_1 + (n-1)d$.

where a_1 is the first term in the sequence, d is the common difference between consecutive terms, and n is the term number.

Using this formula, we have:

$a_n = 7 + (n-1)4$.

Simplifying, we get: $a_n = 4n + 3$.

Therefore, the answer is A.

4) **Choice A is correct**

Considering that for each x_1 and x_2 of the values of the first column of the table of contents of the question such that $x_1 < x_2$, then $f(x_1) < f(x_2)$. So, the correct choice is an increasing function. That is, the base of the exponential function is greater than 1. Finally, by substituting a few points from the first column in the functions of A and B, the answer is obtained. Therefore,

A. $x = 1 \rightarrow f(1) = 24 \left(\frac{3}{2}\right)^1 = 36$, this is true.

B. $x = 1 \rightarrow f(1) = 32 \left(\frac{3}{2}\right)^1 = 48$, this is NOT true.

5) **Choice E is correct**

Since the function h is exponential, it can be written as $h(x) = ab^x$, where a is the y−coordinate of the y−intercept and b is the growth rate. Since it's given that the y−coordinate of the y−intercept is d, the exponential function can be written as $h(x) = db^x$.

6) Choice C is correct

According to the content question, we know that the sum of his hours worked per week is less than 65 hours, it can be shown that her hours worked as $x + y < 65$. First, graph this

equation $x + y = 65$ as a dashed line.

Now, put a point on each side of the dashed line in the inequality and check which one satisfies the inequality. (For example, the points marked in the previous graph.)

A. $(60, 20) \rightarrow 60 + 20 = 80 \not< 65$. It's NOT true!

C. $(20, 20) \rightarrow 20 + 20 = 40 < 65$. It's true.

7) Choice A is correct

First, $y = -x$, then, replace all x's with y and all y's with x: $x = -y$. Now, solve for y: $x = -y \rightarrow -x = y$. Actually, the inverse of $y = -x$ is itself. In other words, the inverse of a graph is asymmetric to $y = x$.

8) Choice C is correct

I. $|a| < 1 \rightarrow -1 < a < 1$

Multiply all sides by b. Since, $b > 0 \rightarrow -b < ba < b$ (it is true!)

II. Since, $-1 < a < 1$, and $a < 0 \rightarrow -a > a^2 > a$ (plug $-\frac{1}{2}$, and check!) (It's false)

III. $-1 < a < 1$, multiply all sides by 2, then: $-2 < 2a < 2$

Subtract 3 from all sides. Then: $-2 - 3 < 2a - 3 < 2 - 3 \rightarrow -5 < 2a - 3 < -1$ (It is true!)

9) Choice D is correct

$m^{\frac{1}{2}} n^{-2} m^4 n^{\frac{2}{3}} \to m^{\frac{1}{2}} \cdot m^4 = m^{\frac{1}{2}+4} = m^{\frac{9}{2}}, \; n^{-2} \cdot n^{\frac{2}{3}} = n^{-2+\frac{2}{3}} = n^{-\frac{4}{3}} = \frac{1}{n^{\frac{4}{3}}}, \; m^{\frac{9}{2}} \cdot \frac{1}{n^{\frac{4}{3}}} = \frac{m^{\frac{9}{2}}}{n^{\frac{4}{3}}}.$

10) Choice C is correct

$\frac{16n^6 - 32n^2 + 8n}{8n} = \frac{16n^6}{8n} - \frac{32n^2}{8n} + \frac{8n}{8n} = 2n^5 - 4n + 1.$

11) Choice C is correct

Check each choice:

A. $2.46 \times 10^1 = 24.6$

B. $2.46 \times 10^2 = 246$

C. $2.46 \times 10^3 = 2,460$

D. $2.46 \times 10^4 = 24,600$

E. $2.46 \times 10^5 = 246,000$

Only 2,460 is between 2,000 and 3,000.

12) Choice C is correct

To solve for $f(3g(P))$, first, find $3g(p)$: $g(x) = log_3 x \to g(p) = log_3 p \to 3g(p) = 3 log_3 p = log_3 p^3$. Now, find $f(3g(p))$: $f(x) = 3^x \to f(log_3 p^3) = 3^{log_3 p^3}$

Logarithms and exponentials with the same base cancel each other. This is true because logarithms and exponentials are inverse operations. Then: $f(log_3 p^3) = 3^{log_3 p^3} = p^3$

13) Choice E is correct

$b^{\frac{m}{n}} = \sqrt[n]{b^m}$ for any positive integers m and n. Thus, $b^{\frac{3}{5}} = \sqrt[5]{b^3}$.

14) Choice A is correct

To find the best representation of the exponential function f, put values of the domain in the equation of the function and compare the producing ordered pairs to the points on the graph. The domain of exponential functions is the set of all real numbers. Therefore, considering some points such as $-1, 0$, and 1, we have:

$x = -1 \to f(-1) = 0.2(25)^{-1} = 0.2 \left(\frac{1}{25}\right) = 0.008 \to (-1, 0.008)$

$x = 0 \to f(0) = 0.2(25)^0 = 0.2(1) = 0.2 \to (0, 0.2)$

$x = 1 \to f(1) = 0.2(25)^1 = 0.2(25) = 5 \to (1,5)$

The generated ordered pairs correspond to the points on the graph of choice A.

15) Choice A is correct

If $f(x) = g(x)$, then: $ln(f(x)) = ln(g(x)) \to ln(e^{2x}) = ln(12)$.

Use the logarithm rule:

$log_a x^b = b\, log_a x \to ln(e^{2x}) = 2x\, ln(e) \to (2x)\, ln(e) = ln(12)$.

Since: $ln(e) = 1$, then: $(2x)\, ln(e) = ln(12) \to 2x = ln(12) \to x = \frac{ln(12)}{2}$.

16) The solution is -5

Simplify, $\frac{2x^2}{5} - 10 = 0$. First, multiply both sides of the equation by 5:

$5 \times \left(\frac{2x^2}{5} - 10\right) = 5 \times 0 \to 5 \times \frac{2x^2}{5} - 5 \times 10 = 0 \to 2x^2 - 50 = 0$.

Add 50 to both sides: $2x^2 - 50 + 50 = 0 + 50 \to 2x^2 = 50$. Now, divide both sides by 2.

Then, $2x^2 \div 2 = 50 \div 2 \to x^2 = 25$. Therefore, $x = \pm 5$.

17) Choice B is correct

To find the correct function for this situation, we can start by determining the base cost for three bags. According to the problem, passengers can check 3 pieces of luggage for the base cost of $150. Therefore, the base cost for three bags is $150.

To find the cost for each additional piece of luggage, we can multiply the number of additional pieces of luggage by the fee of $20. Therefore, the function that represents the total cost in dollars a passenger has to pay to fly with p pieces of luggage, where $b \geq 3$, is:

$c = 20(b - 3) + 150$.

This simplifies to:

$c = 20b + 90$.

Therefore, the answer is B. $c = 20(b - 3) + 150$, which represents the cost in dollars a passenger has to pay to travel with b bags, where $b \geq 3$.

18) Choice E is correct

The function $g(x) = -\frac{1}{8}(x-1)^2 - 3$ is quadratic and the vertex form $y = a(x-h)^2 + k$.

So, there is an axis of symmetry parallel to the y-axis that passes through the vertex. The vertex is the point with the coordinate $(h, k) = (1, -3)$. Therefore, the axis of symmetry is $x = 1$.

19) Choice E is correct

First factor the function: $f(x) = x^3 - 9x^2 + 20x = x(x^2 - 9x + 20) = x(x-4)(x-5)$

To find the zeros, $f(x)$ should be zero. $f(x) = x(x-4)(x-5) = 0$

Therefore, the zeros are: $x = 0$, $(x-4) = 0 \Rightarrow x = 4$, $(x-5) = 0 \Rightarrow x = 5$.

20) Choice D is correct

To find the value of c that makes the graph of h, 12 units above the graph of f, we can set up an equation: $h(x) = f(x) + 12$.

Substitute the given expressions for $f(x)$ and $h(x)$ into this equation:

$3x^2 + c = 3x^2 - 6 + 12$.

Simplify and solve for c: $c = 6$.

Therefore, the value of c that will make the graph of h, 12 units above the graph of f is 6.

21) The value of y is 30

Plug in the value of x in the equation and solve for y. $2y = \frac{2x^2}{3} + 6 \rightarrow 2y = \frac{2(9)^2}{3} + 6 \rightarrow 2y = \frac{2(81)}{3} + 6 \rightarrow 2y = 54 + 6 \rightarrow 2y = 60 \rightarrow y = 30$.

22) Choice C is correct

$-1, 1$, and 3 are the roots of the function so: $x = 1 \rightarrow x - 1 = 0$.

$x = -1 \rightarrow x + 1 = 0$

$x = 3 \rightarrow x - 3 = 0$

Multiply the terms together: $f(x) = (x-1)(x+1)(x-3) \rightarrow f(x) = (x^2 - 1)(x-3)$.

FOIL: $f(x) = (x^2 - 1)(x-3) \rightarrow f(x) = x^3 - 3x^2 - x + 3$.

23) Choice E is correct

The perimeter of the trapezoid is 36 cm. Therefore, the missing side (height) is:

$36 - 8 - 12 - 6 = 10$.

Area of a trapezoid: $A = \frac{1}{2}h(b_1 + b_2) = \frac{1}{2}(10)(6+8) = 70$.

24) The answer is $\frac{10}{3}$

First, factorize the numerator and simplify.

$\frac{x^2-9}{x+3} + 2(x+4) = 15 \to \frac{(x-3)(x+3)}{x+3} + 2x + 8 = 15, \to x - 3 + 2x + 8 = 15 \to 3x + 5 = 15$

Subtract 5 from both sides of the equation. Then: $\to 3x = 15 - 5 = 10 \to x = \frac{10}{3} = 3\frac{1}{3}$.

25) Choice C is correct

Let $x = -3$ and check choices:

A. $2(-3)^2 = 2(9) = 18 < 0$ it's wrong.

B. $\frac{(-3)^2}{8} = \frac{9}{8} < 0$ it's wrong.

C. $(-3)^2 > (-3)^3 \to 9 > -27$ it's true!

D. $\frac{-3}{8} > 0$ it's wrong.

26) Choice E is correct

The function $f(x) = ln(x)$ with domain $x > 0$ and range $-\infty < f(x) < +\infty$ has an x-intercept with coordinates $(1,0)$ where $x = 0$ is a vertical asymptote.

Its graph is as follows:

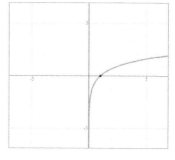

On the other hand, we know that the graph $y = f(x - k); k > 0$, is shifted k units to the right of the graph $y = f(x)$, and if $k < 0$, is shifted k units to the left.

Now, in the example graph, the x-intercept with coordinates $(2,0)$ and vertical asymptote with equation $x = 1$ is shifted 1 unit to the right. Therefore, the function $y = ln(x - 1)$ is represented by an example graph.

27) Choice B is correct

The initial value of the car is $35,000. It's equivalent to the ordered pair $(0, 35000)$. Since the value of the car decreases at a rate of 20%, then the function is decreasing (the choice C is not true) and for the first year $x = 1$, the value of the function is $y = 35,000(0.8)$. Similarly, in the second year $x = 2$, the value of the car is $y = 35,000(0.8)(0.8) = 35,000(0.8)^2$. Therefore, the model of the relationship can be a part of the function of the form $y = 35,000(0.8)^x$ (One of the choices A or B). Now, substitute a few points in the model and check it, until the suitable choice determines.

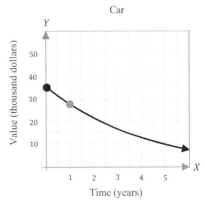

$x = 1 \rightarrow y = 35,000(0.8)^1 = 28,000$. It's equivalent to the value of the graph B.

28) Choice A is correct

The axis of symmetry is a quadratic function, the vertical line along the y-axis (one of the A or C) with the equation $x = a$, where a is the first term of the vertex. The vertex is on the left side of the y-axis. Therefore, the answer is A.

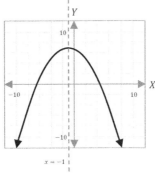

29) Choice D is correct

First, find the value of b, and then find $f(3)$. Since $f(2) = 35$, substuting 2 for x and 35 for $f(x)$ gives $35 = b(2)^2 + 15 \to 35 = 4b + 15$. Solving this equation gives $b = 5$. Thus $f(x) = 5x^2 + 15$, $\quad f(3) = 5(3)^2 + 15 \to f(3) = 45 + 15 \to f(3) = 60$.

30) Choice A is correct

By definition, the x−intercept is the point where a graph crosses the x−axis. If $y = 0$ is put into the equation, it would become:

$y = \frac{x-1}{1-x^2} \to \frac{x-1}{1-x^2} = 0 \to x - 1 = 0 \to x = 1$.

However, we know that the function is not defined for $x = 1$ and $x = -1$, as these values would result in division by zero.

Therefore, the graph of $y = \frac{x-1}{1-x^2}$ would not have x−intercepts.

31) Choice C is correct

Based on the problem information, the late fee for overdue books is $0.2 per day per book. The late fee for 5 overdue books is $1 per day: $5 \times 0.2 = 1$. Therefore, for each late day, $1 will be added to the late fee. The late fee equation for five overdue books is linear and is in the form of $y = x$.

Choice D is false because the maximum late fee is $4.00 per book, which equals $20 for 5 books. Then, the range of the graph related to the late fee of these 5 books is not more than $20.

The choice C is a suitable answer.

32) The maximum value of the graph is 63.4

To find the maximum value of the graph, you need to evaluate the vertex of the quadratic function corresponding to this graph. For this purpose, use the vertex form of a quadratic function as follows:

$$y = a(x - h)^2 + k$$

Where the ordered pair (h, k) is the vertex, and a is a constant number. Now, substitute at least three points on the graph in the vertex form and solve the obtained system of equations. Then:

I. $(0,0) \to 0 = a(0 - h)^2 + k \to ah^2 + k = 0$

II. $(13,0) \to 0 = a(13 - h)^2 + k \to ah^2 - 26ah + 169a + k = 0$

III. $(4,54) \to 54 = a(4 - h)^2 + k \to ah^2 - 8ah + 16a + k = 54$

First, solve the system of equations for $\begin{cases} ah^2 + k = 0 \\ ah^2 - 26ah + 169a + k = 0 \end{cases}$. Subtract the first equation from the second equation. So, $-26ah + 169a = 0 \to a(-26h + 169) = 0 \to a = 0$ or $-26h + 169 = 0 \to 26h = 169 \to h = 6.5$. Since the function is quadratic, then $a \neq 0$. Therefore, $h = 6.5$. Put $h = 6.5$ in these equations and solve. Now, we have this system of equations $\begin{cases} 42.25a + k = 0 \\ 6.25a + k = 54 \end{cases}$ and solve. Then,

$6.25a + k - (42.25a + k) = 54 \to -36a = 54 \to a = -1.5$

Finally, put $a = -1.5$ in the first equation. That is, $42.25(-1.5) + k = 0 \to k = 63.375$. Plug in the values a, h, and k in the vertex form $y = a(x - h)^2 + k$. Therefore,

$$y = -1.5(x - 6.5)^2 + 63.375$$

where $(6.5, 63.375)$ is at the vertex. It means the maximum value of the graph of the function is $k = 63.375 \approx 63.4$

33) Choice E is correct

$f(x) = x^2 + 2x - 10$, $f(4t^2) = (4t^2)^2 + 2(4t^2) - 10 = 16t^4 + 8t^2 - 10$.

34) Choice E is correct

Considering that $f(x)$ opens upward and $g(x)$ opens downward. Rewrite:

$f(x) = 2x^2 - x + 3 \to f(x) = 2\left(x - \frac{1}{4}\right)^2 + \frac{23}{8}$. So: $f(x) \geq \frac{23}{8}$.

On the other hand: $g(x) = -x^2 + 2x + 1 \to g(x) = -(x - 1)^2 + 2 \to g(x) \leq 2$.

Therefore, the minimum of $f(x)$ is greater than the maximum of $g(x)$.

35) Choice E is correct

$y = (-3x^3)^2 = (-3)^2(x^3)^2 = 9x^6$

36) Choice E is correct

Notice that the domain of a quadratic function is all real numbers. That is, the answer is either the choices A or E. In addition, you can see that the graph is going downward. Then the range of this function is less than a real number like a. It is equivalent to the inequality $y \leq a$. The only choice that satisfies these conditions is E.

37) Choice E is correct

The two radical parts are not the same. First, we need to simplify the $2\sqrt{32}$. Then:

$$2\sqrt{32} = 2\sqrt{16 \times 2} = 2(\sqrt{16})(\sqrt{2}) = 8\sqrt{2}.$$

Now, combine like terms:

$$2\sqrt{32} + 2\sqrt{2} = 8\sqrt{2} + 2\sqrt{2} = 10\sqrt{2}.$$

38) Choice D is correct

If $x - a$ is a factor of $g(x)$, then $g(a)$ must equal 0. Based on the table $g(2) = 0$. Therefore,

$x - 2$ must be a factor of $g(x)$.

39) Choice C is correct

To find the best model, it's enough, to substitute the table points in the choices. Any function that satisfies more points is the correct answer. Remember that a quadratic function like $y = ax^2 + bx + c$ is downward if $a < 0$, and vice versa. Therefore, the choices A, B, and E are not suitable answers. By substituting the left column of the table in the remaining functions of the choices, you can see that:

Number of Followers on the Page, x	Number of Feedbacks of each Post for Choice C, y	Number of Feedbacks of each Post for Choice D, y
5	9.625	9.475
10	12.25	11.95
15	16.625	16.175
20	22.75	22.15
25	30.625	29.875
30	40.25	39.35

Of course, it is enough to evaluate and compare a few values. Clearly, the choice C is the best answer.

40) Choice E is correct

At first let's solve with $\Delta = b^2 - 4ac$ so $\Delta = (-6)^2 - 4 \times 1 \times 8 = 4$, the result is positive thus we have two answers: $x = \frac{-b \pm \sqrt{\Delta}}{2a}$ so $x_1 = 4, x_2 = 2$ therefore $a - b = 2$.

41) Choice B is correct

$log_3(x + 20) - log_3(x + 2) = 1$. First, condense the two logarithms:

$log_3(x + 20) - log_3(x + 2) = 1 \to log_3\left(\frac{x+20}{x+2}\right) = 1$.

We know that: $log_a a = 1$. Then:

$log_3\left(\frac{x+20}{x+2}\right) = 1 \to log_3\left(\frac{x+20}{x+2}\right) = log_3 3 \to \frac{x+20}{x+2} = 3$.

Use cross multiplication and solve for x.

$\frac{x+20}{x+2} = 3 \to x + 20 = 3(x + 2) \to x + 20 = 3x + 6 \to 2x = 14 \to x = 7$.

42) Choice E is correct

Simplify. $7x^2y^3(2x^2y)^3 = 7x^2y^3(8x^6y^3) = 56x^8y^6$.

43) Choice C is correct

In the function, $f(2) = 4$. Then: $f(f(2)) = f(4) = 2$.

44) Choice E is correct

First, calculate the product of the multiplication term: $(n + 3)(3n - 5) = 3n^2 + 4n - 15$.

Substitute the obtained term and simplify: $3n^2 + 5n - 3n^2 - 4n + 15 = n + 15$. Then:

$3n^2 + 5n - (n + 3)(3n - 5) = n + 15$.

45) Choice C is correct

The slope of $g(x)$ is $\frac{9}{11}$, while the slope of $h(x)$ is $\frac{3}{2}$. Since the slope is the measure of the steepness of a line, we can see that $g(x)$ has a smaller slope than $h(x)$, which means that it is steeper.

46) Choice A is correct

The absolute value parent function is defined as:
$$f(x) = |x|$$
Now, check each choice:

Option A is incorrect because the graph of the absolute value function is symmetrical about the y−axis, not the x−axis.

Option B is correct for the same reason.

Option C is correct because the domain of the absolute value function is all real numbers.

The absolute value function takes any input value and returns its distance from zero, which is always a non-negative number. Therefore, the range of the absolute value function is all non-negative numbers, and option D is the correct statement about the absolute value parent function.

Option E is correct. Because the graph of the parent function of the absolute value is non-negative for all real numbers. Then this choice is not the suitable solution for this problem.

Option A is a suitable answer.

47) The solution is 2

Let the equation be $P(x) = 6x^2 + 4x - 10$. Factor 2 from the equation: $6x^2 + 4x - 10 = 2(3x^2 + 2x - 5)$. We just need to factor the expression $Q(x) = 3x^2 + 2x - 5$. First, find the roots of the equation $3x^2 + 2x - 5 = 0$ by evaluating the discriminant expression $\Delta = b^2 - 4ac$ of the quadratic equation as $ax^2 + bx + c = 0$. So, we have:
$$\Delta = (2)^2 - 4(3)(-5) = 4 + 60 = 64 \rightarrow \Delta > 0$$
Use the quadratic formula: $x_{1,2} = \frac{-b \pm \sqrt{\Delta}}{2a}$. Therefore, the zeros of the equation $3x^2 + 2x - 5 = 0$ are $x_1 = \frac{-2+\sqrt{64}}{2(3)} = \frac{-2+8}{6} = 1$ or $x_2 = \frac{-2-\sqrt{64}}{2(3)} = \frac{-2-8}{6} = -\frac{10}{6} = -\frac{5}{3}$.

Then, the equation $Q(x) = 0$, can be written in factored form as

$$(x-1)\left(x+\frac{5}{3}\right)=0$$

Multiply both sides of the equation by 3:

$$3\times(x-1)\left(x+\frac{5}{3}\right)=3\times 0 \to (x-1)(3x+5)=0$$

It's equivalent to the equation $3x^2+2x-5=0$. Now, to obtain the expression $P(x)$, multiply both sides of the expression $Q(x)$ by 2. Thus,

$$P(x)=2Q(x) \to P(x)=2(x-1)(3x+5) \to P(x)=(2x-2)(3x+5)$$

Compare the resulting factor with the factored expression in the content of the question:

$$(2x-2)(3x+5)=(2x-m)(3x+5)$$

You can see that the value of m is 2.

48) Choice B is correct

Draw the graph corresponding to each function. Rewrite: $f(x)=2x^2-4x \to f(x)=2(x^2-2x) \to f(x)=2(x-1)^2-2$.

According to a standard form of a parabola, $(1,-2)$ is the vertex. In addition, if $f(x)=0$, then:

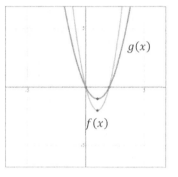

$2x^2-4x=0 \to 2x(x-2)=0 \to x=0, x=2$,

and $(0,0)$ and $(2,0)$ are $x-$intercepts for $f(x)$. Correspondingly, we have for $g(x)$, $(1,-1)$ as its vertex and $(0,0)$, and $(2,0)$ are $x-$intercepts.

Therefore, they have the same $x-$intercepts.

49) Choice C is correct

You can find the possible values of a and b in $(ax+4)(bx+3)$ by using the given equation $a+b=7$ and finding another equation that relates the variables a and b. Since $(ax+4)(bx+3)=10x^2+cx+12$, expand the left side of the equation to obtain:

$abx^2+4bx+3ax+12=10x^2+cx+12$.

Since ab is the coefficient of x^2 on the left side of the equation and 10 is the coefficient of x^2 on the right side of the equation, it must be true that $ab=10$.

The coefficient of x on the left side is $4b + 3a$ and the coefficient of x on the right side is c.

Then: $4b + 3a = c$, $a + b = 7$, then: $a = 7 - b$.

Now, plug in the value of a in equation $ab = 10$. Then:

$ab = 10 \to (7 - b)b = 10 \to 7b - b^2 = 10$.

Add $-7b + b^2$ on both sides. Then: $b^2 - 7b + 10 = 0$.

Solve for b using the factoring method. $b^2 - 7b + 10 = 0 \to (b - 5)(b - 2) = 0$.

Thus, either $b = 2$ and $a = 5$, or $b = 5$ and $a = 2$. If $b = 2$ and $a = 5$, then:

$4b + 3a = c \to 4(2) + 3(5) = c \to c = 23$. If $b = 5$ and $a = 2$, then, $4b + 3a = c \to$

$4(5) + 3(2) = c \to c = 26$. Therefore, the two possible values for c are 23 and 26.

50) Choice E is correct

The sum of the two polynomials is $(4x^2 + 6x - 3) + (3x^2 - 5x + 8)$

This can be rewritten by combining like terms:

$(4x^2 + 6x - 3) + (3x^2 - 5x + 8) = (4x^2 + 3x^2) + (6x - 5x) + (-3 + 8) = 7x^2 + x + 5$.

51) Choice B is correct

We know that the graph of $g(x) = -h(x) = -(3^x)$, is a reflection of the graph of $h(x) = 3^x$ across the x-axis. Also, the graph of $f(x) = 4h(x) = 4(3^x)$, is a vertical stretch because the number 4 is positive. Now, combining f, g, and h creates the function k:

$$f \circ g \circ h(x) = k(x) = -4(3^x)$$

Then, the choice B is the correct answer.

52) Choice B is correct

To find the value of $f(8)$, make $x^3 = 8$. Then, $x^3 = 8 \to x = 2$. Finally,

$$x = 2 \to f(2^3) = 2(2) - 5 = -1 \to f(8) = -1.$$

53) Choice C is correct

The graph is upward. It means the coefficient of the leading term is positive. Therefore, the choices D and E are not the answer. According to the graph, you can see that the function has two zeros at -1 and 3. Since the points on the graph must be true in the corresponding equation. Then, plug $x = -1$ into the choices and

A. $y = (-1)^2 + 2(-1) - 3 = -4 \neq 0$

B. $y = (-1)^2 - 2(-1) + 3 = 6 \neq 0$

C. $y = (-1)^2 - 2(-1) - 3 = 0$

The choice C is the best function.

54) Choice B is correct

Notice that $(bx - a)$ is a factor of an expression whenever $x = \frac{a}{b}$ is a zero of that expression. Check which of the choices is a zero of the expression.

A. $x + 1 = 0 \rightarrow x = -1 \rightarrow 15(-1)^2 - 12(-1) - 3 = 15 + 12 - 3 = 24 \neq 0$

B. $5x + 1 = 0 \rightarrow x = -\frac{1}{5} \rightarrow 15\left(-\frac{1}{5}\right)^2 - 12\left(-\frac{1}{5}\right) - 3 = 15\left(\frac{1}{25}\right) + \frac{12}{5} - 3 = \frac{15+60-75}{25} = 0$

C. $5x - 1 = 0 \rightarrow x = \frac{1}{5} \rightarrow 15\left(\frac{1}{5}\right)^2 - 12\left(\frac{1}{5}\right) - 3 = 15\left(\frac{1}{25}\right) - \frac{12}{5} - 3 = \frac{15-60-75}{25} = -\frac{24}{5} \neq 0$

D. $3x - 1 = 0 \rightarrow x = \frac{1}{3} \rightarrow 15\left(\frac{1}{3}\right)^2 - 12\left(\frac{1}{3}\right) - 3 = 15\left(\frac{1}{9}\right) - \frac{12}{3} - 3 = \frac{5-12-9}{3} = -\frac{16}{3} \neq 0$

E. $3x + 1 = 0 \rightarrow x = -\frac{1}{3} \rightarrow 15\left(-\frac{1}{3}\right)^2 - 12\left(-\frac{1}{3}\right) - 3 = 15\left(\frac{1}{9}\right) + \frac{12}{3} - 3 = \frac{5+12-9}{3} = \frac{8}{3} \neq 0$

Therefore, B is the correct answer.

55) Choice C is correct

Note that a set represents a function if, for every point in the domain, there is at most one point in the range. Check the table information in each choice as follows:

A. This is not a function because there is more than one value for the -2 point in the domain.

B. Since there are two values 3 and 10 for 1, it is not a function.

C. This table shows the information of a function.

D. This is not a function, because the -17 of the domain corresponds to the two values 0 and -4 of the range.

E. This is not a function, because the 0 of the domain corresponds to the two values 0 and -2 of the range.

Therefore, the choice C is the correct answer.

56) Choice A is correct

First, determine the common factors from each part of the expression: $35r^2 = 7 \times 5 \times r \times r$ and $28r = 7 \times 2 \times 2 \times r$. The common factors are 7 and r. Now, Rewrite as follows:

$$35r^2 - 28r = 7r(5r) - 7r(4)$$

Factor in $7r$: It means that $7r(5r) - 7r(4) = 7r(5r - 4)$. Therefore,

$$35r^2 - 28r = 7r(5r - 4).$$

57) Choice E is correct

We know that the factors of an expression $p(x)$ are the zeros of the equation $p(x) = 0$. Find the zeros of the quadratic equation as follows. First, evaluate Δ from the formula $\Delta = b^2 - 4ac$ for the quadratic equation like $ax^2 + bx + c = 0$. Then:

$$\Delta = (5)^2 - 4(25)(1) = 25 - 100 = -75 < 0$$

There are no real solutions. Therefore, there are no real factors.

58) Choice B is correct

The graph is a descending exponential function, so the base of the corresponding exponential function is a positive number smaller than 1 (One of the choices A, B, or C). Now, by substituting some points on the graph in the possible choices, the equation in which the points of the graph satisfy is the answer to the question. We check point $(1,2)$. So, we have:

A. $(1,2) \to p(1) = 4(0.25)^1 = 4(0.25) = 1 \neq 2$. It is NOT true!

B. $(1,2) \to p(1) = 4(0.5)^1 = 4(0.5) = 2$. This is the suitable answer.

C. $(1,2) \to p(1) = (0.75)^1 = (0.75) = 0.75 \neq 2$. It is NOT true!

59) Choice B is correct

To rewrite $\frac{2+3i}{5-2i}$ in the standard form $a+bi$, multiply the numerator and denominator of $\frac{2+3i}{5-2i}$ by the conjugate, $5+2i$. This gives $\left(\frac{2+3i}{5-2i}\right)\left(\frac{5+2i}{5+2i}\right) = \frac{10+4i+15i+6i^2}{5^2-(2i)^2}$. Since $i^2 = -1$, this last fraction can be rewritten as $\frac{10+4i+15i+6(-1)}{25-4(-1)} = \frac{4+19i}{29}$.

60) Choice A is correct

The standard form equation of a parabola is $y = ax^2 + bx + c$.

$y = (x+3)(x-9) \to y = x^2 - 6x - 27$.

So, the $x-$coordinate of the vertex is $\frac{-b}{2a} = \frac{-(-6)}{2(1)} = 3$.

Substituting 3 in the original equation to get the $y-$coordinate, we get:

$y = 3^2 - 6(3) - 27 = -36$.

High School Algebra I Practice Tests 2 Explanations

1) **Choice A is correct**

According to the table, the value of y increases by $\frac{21}{10} - \frac{8}{5} = \frac{21-16}{10} = \frac{5}{10} = \frac{1}{2}$ every time the value of x increases by 1. It follows that the simplest equation relating y to x is linear and of the form $y = \frac{1}{2}x + b$ for some constant b. Furthermore, the ordered pair $\left(2, \frac{8}{5}\right)$ from the table must satisfy this equation. Substituting 2 for x and $\frac{8}{5}$ for y in the equation $y = \frac{1}{2}x + b$ gives $\frac{8}{5} = \frac{1}{2}(2) + b$. Solving this equation for b gives $b = \frac{3}{5}$.

Therefore, the equation in choice A correctly relates y to x.

2) **Choice C is correct**

There can be 0, 1, or 2 solutions to a quadratic equation. In standard form, a quadratic equation is written as $ax^2 + bx + c = 0$.

For the quadratic equation, the expression $b^2 - 4ac$ is called the discriminant. If the discriminant is positive, there are 2 distinct solutions for the quadratic equation. If the discriminant is 0, there is one solution for the quadratic equation and if it is negative the equation does not have any solutions.

To find the number of solutions for $x^2 = 4x - 3$, first, rewrite it as $x^2 - 4x + 3 = 0$.

Find the value of the discriminant. $b^2 - 4ac = (-4)^2 - 4(1)(3) = 16 - 12 = 4$.

Since the discriminant is positive, the quadratic equation has two distinct solutions.

3) **Choice A is correct**

A set represents a function if for every input there is at most one point in the output. By checking each choice, you see that choice A is NOT a function. For point 2, there are two values -8 and 8 for the function.

4) **Choice C is correct**

Considering the choices, the corresponding function of this graph is exponential. Since the graph is increasing, then the suitable choice has a base greater than 1. That is, the choices A and B are not the answer. Now, by substituting some points on the graph with the remaining choices and checking to satisfy these points, the correct answer is determined.

Try the points (0,32). Therefore,

C. $(0,32) \to 32(1.5)^0 = 32$. It's true.

D. $(0,32) \to 1.5(32)^0 = 1.5 \neq 32$. It's NOT true!

E. $(0,32) \to 0.67(48)^0 = 0.67 \neq 32$. It's NOT true!

Then, the choice C is the correct answer.

5) The answer is -3

$g(f(1)) = g(5 + c) = 14 + 4c$ and $g(2) = 2$. So, $g(f(1)) \times g(2) = 4$
$\to (14 + 4c)(2) = 4 \to 28 + 8c = 4 \to 8c = -24 \to c = -3$

6) Choice E is correct

Use the FOIL (First, Out, In, Last) method:

$(5x + 2y)(2x - y) = 10x^2 - 5xy + 4xy - 2y^2 = 10x^2 - xy - 2y^2$.

7) Choice C is correct

According to the first characteristic, one of the choices of either B or C can be the right choice. Since the graph has two zeros, then the points where the graph intersects the x −axis are two points. Therefore, only choice C is correct. The graph below shows the desired characteristics of choice C.

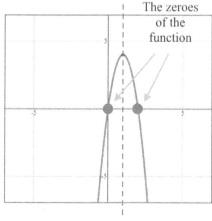

The axis of symmetry
$x = 1$

8) Choice D is correct

The equation of the quadratic function $y = 3x^2 - 6x$ is in the standard form $y = ax^2 + bx + c$. So, the axis of symmetry is obtained from the formula $x = \frac{-b}{2a}$. Therefore, the axis of symmetry of this function is $x = \frac{-(-6)}{2(3)} \to x = 1$. The function is upward and has one minimum point which is equivalent to the vertex.

Factor as follow: $3x^2 - 6x = 3x(x) - 3x(2) = 3x(x - 2)$. That is, the factors of g are $3x$ and $x - 2$. By calculating $g(x) = 0$, the zeros of the equation are 0 and 2. Because,

$$g(x) = 0 \to 3x(x - 2) = 0 \to x = 0 \text{ and } x = 2.$$

For quadratic functions, there is no horizontal axis of symmetry with the equation $y = ax + b$. Then, the choice D is correct.

9) Choice C is correct

To find the correct ordered pair, the choices can be put into inequality. Whichever choice applies to the inequality is the correct answer.

A. $(0,3) \to 3 - \frac{5}{2}(0) = 3 \not< 3$

B. $(0,-6) \to 3 - \frac{5}{2}(0) = 3 \not< -6$

C. $(6,-4.5) \to 3 - \frac{5}{2}(6) = -12 < -4.5$

D. $(-1,0) \to 3 - \frac{5}{2}(-1) = 5.5 \not< 0$

E. $(0,0) \to 3 - \frac{5}{2}(0) = 3 \not< 0$

The choice C is the answer.

10) Choice D is correct

First, factor the function: $f(x) = x^3 + 7x^2 + 12x = x(x^2 + 7x + 12) = x(x + 3)(x + 4)$.

To find the zeros, $f(x)$ should be zero: $f(x) = x(x + 3)(x + 4) = 0$.

Therefore, the zeros are: $x = 0, (x + 3) = 0 \Rightarrow x = -3, (x + 4) = 0 \Rightarrow x = -4$.

11) Choice E is correct

Since $(0, 0)$ is a solution to the system of inequalities, substituting 0 for x and 0 for y in the given system must result in two true inequalities. After this substitution, $y < c - x$ becomes $0 < c$, and $y > x + b$ becomes $0 > b$. Hence, c is positive and b is negative. Therefore, $c > b$.

12) Choice A is correct

Let x be the number of years. Therefore, $\$2,000$ per year equals $2,000x$. Starting from a $\$26,000$ annual salary means you should add that amount to $2,000x$. Income more than that is: $I > 2,000x + 26,000$.

13) Choice D is correct

For the values of x that make the denominator zero, the function is not defined.
$x^2 - x - 2 = 0 \rightarrow (x + 1)(x - 2) = 0 \rightarrow x = -1$ or $x = 2$

14) Choice C is correct

First, find the factors of the numerator and denominator of the expression. Then simplify.
$$\frac{5x^2+75x-80}{x^2-1} = \frac{5(x^2+15x-16)}{(x-1)(x+1)} = \frac{5(x+16)(x-1)}{(x-1)(x+1)} = \frac{5(x+16)}{(x+1)} = \frac{5x+80}{x+1}$$

15) Choice A is correct

If $r = 55 \rightarrow \frac{x^2+6x-55}{x-5} = \frac{(x+11)(x-5)}{(x-5)} = x + 11$, for all other options, the numerator expression is not divisible by $(x - 5)$.

16) Choice A is correct

Since we know that a quadratic function as $y = ax^2 + bx + c$ is upward if $a > 0$, and vice versa. So, the choices B, C and D are not the suitable answer. Now, by substituting a few points in the remaining functions, we select the best quadratic model.

x	The choice A	The choice E
5.7	83.01	121.674
5.9	109.15	140.566
6	120.4	148.2
6.1	130.45	154.626
6.5	158.55	168.25
6.95	167.06	160.474
7.01	166.35	157.58896
7.8	116.4	79.044
8.3	45.83	−9.626

According to the data in the recent table, A is clearly the best answer.

17) Choice A is correct

$2x - 5 \geq 3x - 1$, Add 5 to both sides: $2x - 5 + 5 \geq 3x - 1 + 5 \rightarrow 2x \geq 3x + 4$, subtract $3x$ from both sides: $2x - 3x \geq 3x + 4 - 3x \rightarrow -x \geq 4$. Multiply both sides by -1 (reverse the inequality):

$(-x)(-1) \geq (4)(-1) \rightarrow x \leq -4$. Only -2 is greater than -4.

18) Choice E is correct

Remember that the range of an arbitrary function is an image of the graph on the y−axis. We know that the ordered pair $(2,4)$ is not on the graph and the range of an exponential function is the interval $(0, +\infty)$ that is, the point 0 is not in the range of an exponential function. So, the image of the graph on the y−axis covered the interval $0 < y < 4$.

19) Choice A is correct

$12x^4 + n = a(x^2 + 3)(x^2 - 3) = ax^4 - 9a \rightarrow a = 12$ And $n = -9a = -9 \times 12 = -108$.

20) Choice B is correct

Use the FOIL (First-Out-In-Last) method:

$-(2-i)(i-2) = -2i + 4 + i^2 - 2i = -4i + i^2 + 4$.

Combine like terms:

$(2i - 3) - 4i + i^2 + 4 = -2i + i^2 + 1 = -2i + (-1) + 1 = -2i$.

21) Choice E is correct

$x^2 = 121 \rightarrow x = 11$ (Positive value) or $x = -11$ (negative value).

Since x is positive, then: $f(121) = f(11^2) = 3(11) + 4 = 33 + 4 = 37$.

22) Choice E is correct

The domain of motion of this throw, from the starting point to the point entering the basketball net, is equivalent to the changes of the first component of the ordered pairs. It means that the interval $0 \leq x \leq 14.5$.

23) Choice E is correct

Since the expression is equivalent to nx^3, we get:

$(5m + 5)x^3 - (m^2 - 16)x = nx^3 \rightarrow \begin{matrix} 5m + 5 = n \\ m^2 - 16 = 0 \end{matrix} \rightarrow m = 4$ or $m = -4$

m is positive. So, m is 4.

To find the value of n, substitute the value of $m = 4$ to the equation $n = 5m + m$. Therefore,

$n = 5 \times 4 + 5 = 25$.

The choice E is the correct answer.

24) Choice B is correct

To solve this problem, first recall the equation of a line: $y = mx + b$, where $m =$ slope, $y = y-$intercept. Remember that slope is the rate of change that occurs in a function and that the $y-$intercept is the y value corresponding to $x = 0$. Since the height of John's plant is 5 inches tall when he gets it. Time (or x) is zero. The plant grows 3 inches per year. Therefore, the rate of change of the plant's height is 3. The $y-$intercept represents the starting height of the plant which is 5 inches.

25) Choice E is correct

Since $y = f(x)$, the value of $f(0)$ is equal to the value of $f(x)$, or y, when $x = 0$. The graph indicates that when $x = 0, y = 2$. It follows that the value of $f(0) = 2$.

26) Choice E is correct

The smallest $y-$coordinate belongs to the point with coordinates $(3, -4)$. The minimum value of the graph is $f(3) = -4$. Therefore, the value of $f(x)$ is at its minimum when x equals 3.

27) Choice C is correct

The slope of both graphs is positive. So, to make graph g, the slope of graph f must be multiplied by a positive number. In addition, the slope of the graph g is steeper than the original graph. Therefore, the slope of the graph g is multiplied by a number greater than one. On the other hand, we can see that graph g is 4 units higher than the graph f. The only choice that meets these conditions is option C.

28) Choice B is correct

To solve the system of equations:
$$3x + y = 15$$
$$-2x + 4y = 10$$

We can use the method of elimination. We want to eliminate one of the variables, so we can start by eliminating x by multiplying the second equation by $\frac{3}{2}$ to get:

$$\begin{array}{c} 3x + y = 15 \\ \frac{3}{2} \times (-2x + 4y) = \frac{3}{2} \times (10) \end{array} \rightarrow \begin{array}{c} 3x + y = 15 \\ -3x + 6y = 15 \end{array}$$

Next, we can add the two equations to eliminate y and solve for x:

$3x + y + (-3x + 6y) = 15 + 15 \rightarrow 7y = 30 \rightarrow y = \frac{30}{7}$.

Now, substitute $y = \frac{30}{7}$, in the first equation and solve:

$3x + \left(\frac{30}{7}\right) = 15 \rightarrow 3x = 15 - \frac{30}{7} \rightarrow 3x = \frac{75}{7}$

$$\rightarrow x = \frac{25}{7}$$

Therefore, the solution of the system of equations is $\left(\frac{25}{7}, \frac{30}{7}\right)$, which means the answer is option B.

29) Choice B is correct

Line AB is the best-fit line. Then, point $(6,1)$ is the farthest from line AB.

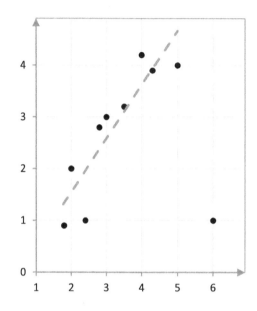

30) Choice C is correct

To find the recursive formula, start by looking at the common differences and ratios of consecutive terms. By evaluating the difference between terms with the previous term, you notice that the differences between consecutive terms are all the same. That is, $a_n - a_{n-1} = 4$. In this step, for calculating the first term of the arithmetic sequence, substitute one of the given terms like $a_4 = 19$ and the common difference $d = 4$ in the arithmetic sequence formula: $a_n = a_1 + d(n-1)$, where $a_1 =$ the first term, $d =$ the common difference between terms, $n =$ number of items. Then,

$$a_4 = a_1 + 4(4-1) = 19 \to a_1 + 12 = 19 \to a_1 = 7$$

Therefore, the nth term of the sequence is $a_n = 7 + 4(n-1)$. Now, simplify as

$$a_n = 4n + 3.$$

31) Choice E is correct

Bring all values to one side of the equation. Add x to both sides:

$$3x^2 - 4x + x = 3 - x + x \to 3x^2 - 3x = 3.$$

Subtract 3 from both sides:

$$3x^2 - 3x - 3 = 3 - 3 \to 3x^2 - 3x - 3 = 0.$$

Now, solve using the quadratic formula: For a quadratic form $ax^2 + bx + c = 0$, the solutions are $x_{1,2} = \frac{-b \pm \sqrt{b^2 - 4ac}}{2a}$. For this equation: $a = 3, b = -3, c = -3$. Then:

$$x_{1,2} = \frac{-(-3) \pm \sqrt{(-3)^2 - 4(3)(-3)}}{2(3)}$$

$x_1 = \frac{-(-3) + \sqrt{(-3)^2 - 4(3)(-3)}}{2(3)} = \frac{3 + \sqrt{45}}{6} = \frac{3 + 3\sqrt{5}}{6}$. Cancel the common factor 3 $\to = \frac{1 + \sqrt{5}}{2}$.

$x_2 = \frac{-(-3) - \sqrt{(-3)^2 - 4(3)(-3)}}{2(3)} = \frac{3 - \sqrt{45}}{6} = \frac{3 - 3\sqrt{5}}{6}$. Cancel the common factor 3 $\to = \frac{1 - \sqrt{5}}{2}$.

32) Choice A is correct

Since $f(x)$ is a linear function with a negative slope, then when $x = -2$, $f(x)$ is maximum, and when $x = 3$, $f(x)$ is minimum. Then the ratio of the minimum value to the maximum value of the function is $\frac{f(3)}{f(-2)} = \frac{-3(3) + 1}{-3(-2) + 1} = \frac{-8}{7} = -\frac{8}{7}$.

33) The answer is 750

The inequalities $y \leq -15x + 3{,}000$ and $y \leq 5x$ can be graphed in the xy-plane. They are represented by the lower half-planes with the boundary lines $y = -15x + 3{,}000$ and $y = 5x$, respectively. The solution set of the system of inequalities will be the intersection of these half-planes, including the boundary lines, and the solution (a, b) with the greatest possible value of b will be the point of intersection of the boundary lines. The intersection of boundary lines of these inequalities can be found by substituting $5x$ for y in the equation for the first line:

$5x = -15x + 3{,}000$, which has solution $x = 150$. Thus, the x-coordinate of the point of intersection is 150. Therefore, the y-coordinate of the point of intersection of the boundary lines is $y = 5(150) = -15(150) + 3{,}000 = 750$. This is the maximum possible value of b for a point $(a, b) = (150, 750)$ that is in the solution set of the system of inequalities.

34) Choice A is correct

Since the data model is exponential and the data is decaying, therefore the base of the exponential function is less than one. It means that one of the choices A or B is the answer. Now, plug a few points from the left column likes 1 or even 2 in the choices to determine the best model. So,

A. $x = 1 \rightarrow n(1) = 71,235(0.77)^1 = 54,850.95$

$x = 2 \rightarrow n(2) = 71,235(0.77)^2 = 42,235.23$

B. $x = 1 \rightarrow n(1) = 71,235(0.89)^1 = 63,399.15$

$x = 2 \rightarrow n(2) = 71,235(0.89)^2 = 56,425.24$

Clearly, the choice A is the best example of the data.

35) Choice E is correct

The area of the triangle is: $\frac{1}{2} AD \times BC$ and AD is perpendicular to BC.

Triangle ADC is a $30° - 60° - 90°$ right triangle.

The relationship among all sides of the right triangle $30° - 60° - 90°$ is provided in the following triangle: In this triangle, the opposite side of the $30°$ angle is half of the hypotenuse. And the opposite side of $60°$ is opposite of $30° \times \sqrt{3}$

$CD = 6$, then $AD = 6 \times \sqrt{3}$

Area of the triangle ABC is: $\frac{1}{2} AD \times BC = \frac{1}{2}(6\sqrt{3}) \times 12 = 36\sqrt{3}$

36) Choice E is correct

You must factor in each of the quadratics to solve this problem.

$x^2 - 3x - 10 = 0 \rightarrow (x - 5)(x + 2) = 0$. Find the zeroes: $x = 5$ or $x = -2$

$x^2 - x - 20 = 0 \rightarrow (x - 5)(x + 4) = 0$. Find the zeroes: $x = 5$ or $x = -4$

Since both equations are true, then x must be 5.

37) Choice C is correct

Each additional year of age over 15 years adds $1.3\ ft$ to the average height of this type of oak tree. The equation $h = 1.3(a - 15) + 45$ represents the relationship between the average height h and the age of the tree a. The coefficient 1.3 in the equation tells us that for each additional year of age over 15 years, the average height increases by 1.3 feet. Therefore, option C is correct.

38) Choice E is correct

Plug in the values of x and y in the equation of the parabola. Then:

$12 = a(2)^2 + 5(2) + 10 \to 12 = 4a + 10 + 10 \to 12 = 4a + 20$

$\to 4a = 12 - 20 = -8 \to a = \frac{-8}{4} = -2 \to a^2 = (-2)^2 = 4.$

39) Choice A is correct

The parent graph of the given function $y = -x^2 + 2x$, is $y = x^2$. The parent graph has undergone the following transformations:

- The coefficient of x^2 is -1, indicating that the graph is reflected across the $x-$axis.
- The constant term, $2x$, indicates that the graph is shifted 1 unit to the right and 1 unit up.

Therefore, the correct answer is A. The parent graph is $y = x^2$, which is reflected across the $x-$axis and shifted 1 unit to the right and 1 unit up.

40) The $y-$intercept is 11.5

The $y-$intercept in a graph is the value that intercepts the $y-$axis function. For this purpose, calculate the value of the function for $x = 0$. Therefore,

$x = 0 \to f(0) = 11.5(0.8)^0 = 11.5 \times 1 = 11.5$

41) Choice B is correct

$0.00035 \times (1.2 \times 10^4) = (3.5 \times 10^{-4}) \times (1.2 \times 10^4) = (3.5 \times 1.2) \times (10^{-4} \times 10^4)$

$= 4.2 \times 10^{-4+4} = 4.2$

42) The solution is 3

Use the exponential rule: $(x^a)^b = x^{a \times b}$. Then, $(n^2)^3 = n^{2 \times 3} = n^6$. Substitute n^6 in $n^{-3}(n^2)^3$:

$$n^{-3}(n^2)^3 = n^{-3}n^6$$

By using the rule $x^a \times x^b = x^{a+b}$, we have: $n^{-3}n^6 = n^{-3+6} = n^3$. Compare n^3 with n^x. So, the value of x is 3.

43) Choice D is correct

Use this polynomial identity: $x^2 + (a+b)x + ab = (x+a)(x+b)$. Considering the given equation $k^2 - 17k + 66$. We need to find two numbers whose product is 66 and their sum is -17. According to the given choices, these two numbers are -6 and -11 for the above formula as follows:

$$k^2 - 17k + 66 = (k-6)(k-11)$$

44) Choice A is correct

To solve the problem, we can use the formula for the perimeter of a rectangle, which is:

Perimeter $= 2 \times$ Length $+ 2 \times$ Width.

We are given that the perimeter of the rectangle is 38 centimeters. We are also given that the length can be represented as $(x+6)$ and the width can be represented as $(2x-2)$. So, we can substitute these values into the formula for the perimeter and solve for x:

$38 = 2(x+6) + 2(2x-2) \to 38 = 2x + 12 + 4x - 4 \to 38 = 6x + 8$

$\to 30 = 6x \to x = 5.$

Now, we can find the length and width of the rectangle by substituting $x = 5$ into the expressions for the length and width:

Length $= x + 6 = 5 + 6 = 11\ m$, and Width $= 2x - 2 = 2(5) - 2 = 8\ m$.

Therefore, the dimensions of the rectangle are $11\ m$ by $8\ m$.

45) Choice C is correct

Multiply the numerators and denominators: $\frac{3x+6}{x+5} \times \frac{x+5}{x+2} = \frac{(3x+6)(x+5)}{(x+5)(x+2)}$

Cancel the common factor: $\frac{(3x+6)(x+5)}{(x+5)(x+2)} = \frac{(3x+6)}{(x+2)}$

Factor $3x + 6 = 3(x+2)$

Then: $\frac{3(x+2)}{(x+2)} = 3.$

46) Choice C is correct

If there is a line parallel to the y-axis that intersects the graph at more than one point. Then that graph does not represent a function. Clearly, the graph of choices A, B, and D shows the graph of a straight line, a quadratic equation, and an exponential. Then these graphs are functions.

In the case of choice C, the line shown in the graph intersects the graph at more than one point, so it is not a function.

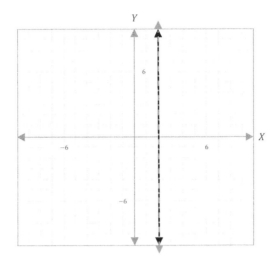

47) Choice A is correct

We know that for the function f, if $0 < k < 1$ then the function is compressed vertically.

It's equivalent to graph A.

48) Choice A is correct

The formula for the surface area of a cylinder is:

$SA = 2\pi r^2 + 2\pi rh \rightarrow 150\pi = 2\pi r^2 + 2\pi r(10)$

Both sides divided by 2π: $\rightarrow r^2 + 10r - 75 = 0$

$(r + 15)(r - 5) = 0 \rightarrow r = 5$ or $r = -15$ (unacceptable)

49) Choice E is correct

Reflect the function f about the x-axis. Multiply the function f by -1. So, there exists a function like k such that $k(x) = -f(x)$. See the graph below,

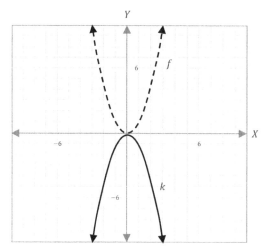

By shifting k two units to the right, the function g is obtained. Use the property such that the function $y = f(x - k)$ shifted to the right, if $k > 0$. Therefore, $g(x) = k(x - 2)$. We know that,

$$k(x) = -f(x) \rightarrow k(x - 2) = -f(x - 2)$$

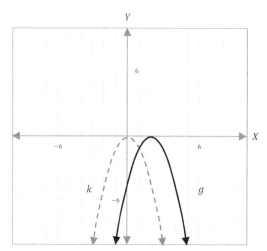

Substitute $k(x - 2) = -f(x - 2)$ in the function $g(x) = k(x - 2)$ and simplify:

$$g(x) = k(x - 2) \rightarrow g(x) = -f(x - 2)$$

If the function f is the parent function equal to $f(x) = x^2$, then:

$$f(x) = x^2 \rightarrow g(x) = -f(x - 2) = -(x - 2)^2.$$

50) Choice E is correct

Multiply by the conjugate: $\frac{\sqrt{12}+3}{\sqrt{12}+3} \to \frac{6}{\sqrt{12}-3} \times \frac{\sqrt{12}+3}{\sqrt{12}+3}$

$(\sqrt{12}-3)(\sqrt{12}+3) = 3$, then: $\frac{6}{\sqrt{12}-3} \times \frac{\sqrt{12}+3}{\sqrt{12}+3} = \frac{6(\sqrt{12}+3)}{3} = 2(\sqrt{12}+3)$.

51) The answer is $\frac{4}{7}$

$\frac{8+3x}{7} - \frac{4-4x}{7} = \frac{8+3x-4+4x}{7} = \frac{7x+4}{7} = \frac{7x}{7} + \frac{4}{7} = x + \frac{4}{7}$. Therefore, this expression is $\frac{4}{7}$ greater than x.

52) Choice D is correct

y is the intersection of the three circles. Therefore, it must be odd (from circle A), negative (from circle B), and multiple of 7 (from circle C). From the options, only -21 is odd, negative, and multiple of 7.

53) Choice B is correct

The input value is 5. Then: $x = 5$. $f(x) = x^2 - 3x \to f(5) = 5^2 - 3(5) = 25 - 15 = 10$.

54) Choice E is correct

The problem asks for the sum of the roots of the quadratic equation $2n^2 + 16n + 24 = 0$. Dividing each side of the equation by 2 gives $n^2 + 8n + 12 = 0$. If the roots of $n^2 + 8n + 12 = 0$ are n_1 and n_2, then the equation can be factored as $n^2 + 8n + 12 = (n - n_1)(n - n_2) = 0$. Looking at the coefficient of n on each side of $n^2 + 8n + 12 = (n + 6)(n + 2)$ gives $n = -6$ or $n = -2$, then, $-6 + (-2) = -8$.

55) The solution is $\frac{7}{2}$

First, simplify the equation $6(x-1)^2 = 41 - x$. Expand the square of the binomial: $(x-1)^2 = (x^2 - 2x + 1)$. Substitute the binomial in the equation

$$6(x^2 - 2x + 1) = 41 - x$$

Then,

$$6x^2 - 12x + 6 = 41 - x$$

Subtract 41 from both sides: $6x^2 - 12x + 6 - 41 = -x \to 6x^2 - 12x - 35 = -x$. Add x to both sides: $6x^2 - 12x - 35 + x = -x + x \to 6x^2 - 11x - 35 = 0$.

Now, evaluate the discriminant of the quadratic equation $6x^2 - 11x - 35 = 0$. The expression $\Delta = b^2 - 4ac$ is called the discriminant for the standard form of the quadratic equation as $ax^2 + bx + c = 0$. So, $\Delta = b^2 - 4ac \rightarrow \Delta = (-11)^2 - 4(6)(-35) = 961 > 0$. Then, the quadratic equation has two distinct solutions. Use the formula:

$$x_{1,2} = \frac{-b \pm \sqrt{\Delta}}{2a}$$

Therefore, the roots are $x_1 = \frac{-(-11)+\sqrt{961}}{2(6)} = \frac{11+31}{12} = \frac{7}{2}$, and $x_2 = \frac{-(-11)-\sqrt{961}}{2(6)} = \frac{11-31}{12} = -\frac{5}{3}$.

56) Choice B is correct

To solve for the inverse function, first, replace $f(x)$ with y. Then, solve the equation for x and after that replace every x with a y and replace every y with an x. Finally, replace y with $f^{-1}(x)$.

$$f(x) = \frac{10x - 3}{6} \Rightarrow y = \frac{10x - 3}{6} \Rightarrow 6y = 10x - 3 \Rightarrow 6y + 3 = 10x \Rightarrow \frac{6y + 3}{10} = x$$

$$f^{-1}(x) = \frac{6x + 3}{10} \Rightarrow f^{-1}(2) = \frac{6(2) + 3}{10} = \frac{15}{10} = \frac{3}{2}$$

57) Choice D is correct

Since $4x = \frac{48}{3}$ and 48 divided by 3 is 16, which gives equation $4x = 16$, then dividing both sides of $4x = 16$ by 4 gives $x = 4$. Therefore $x - 2 = 4 - 2 = 2$ and 7 to the power of 2 is 49. Choices A, B, C and E are incorrect.

58) Choice A is correct

Since none of the given points in the set $\{(2,3)(0,-1)(4,7)(-1,-3)(-3,-7)\}$ are related by any exponent or root, we can eliminate options B, C, and D, leaving only option A as a possible parent function. Therefore, the parent function of f is $y = x$.

59) Choice E is correct

To find the y-intercept, put the value of x in the function $g(x)$ equal to zero. Therefore, $x = 0 \rightarrow g(0) = 25(1.2)^{0+1} = 25(1.2) = 30$.

60) The y-intercept is -2

The y-intercept is the point where the graph intersects the y-axis. The point is shown in the graph below.

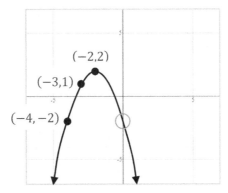

The above quadratic function is symmetric with respect to the axis $x = -2$. So, the ordered pair $(0, -2)$ is the y-intercept.

Receive the PDF version of this book or get another FREE book!

Thank you for using our Book!

Do you LOVE this book?

Then, you can get the PDF version of this book or another book absolutely FREE!

Please email us at:

info@EffortlessMath.com

for details.

Author's Final Note

Congratulations on completing "High School Algebra I"! Your dedication and determination have brought you to the end, and I commend you for your effort.

With numerous study materials at your fingertips, it speaks volumes that you chose this guide to accompany you on your Algebra I journey. I truly appreciate your trust in this book as your academic ally.

Compiling this guide was an endeavor born out of my passion for making Algebra I accessible and comprehensible for all. Over the years, I've distilled my teaching experiences, student feedback, and extensive research into this comprehensive guide, all to provide you with the best possible resource for your studies.

For any questions, uncertainties, or feedback, please don't hesitate to reach out at reza@effortlessmath.com. Your insights are invaluable in refining future editions of this book. And, in the tradition of mathematics, should you identify any discrepancies or areas of improvement, please alert me so that they can be rectified promptly.

If this book has played a part in bolstering your understanding of Algebra I and contributed positively to your academic pursuits, I'd be overjoyed to hear from you. Kindly consider sharing your experiences by leaving a review on the book's Amazon page.

I closely monitor each review, seeking ways to elevate the learning experience for every reader. By leaving a review, you not only assist me but countless other Algebra students who will benefit from your insights.

Wishing you all the success in your future mathematical endeavors and beyond!

With gratitude,

Reza Nazari

Math teacher and author

Made in the USA
Las Vegas, NV
07 June 2025

23319820R00216